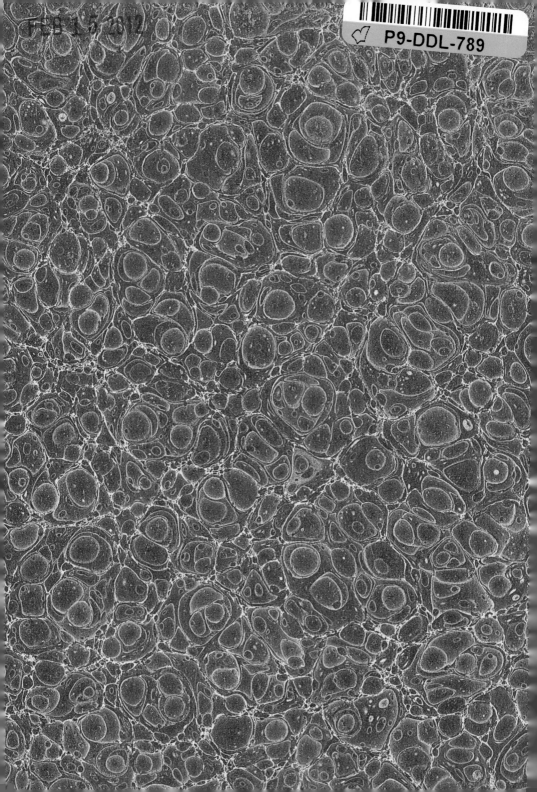

FEB 15 2012

P9-DDL-789

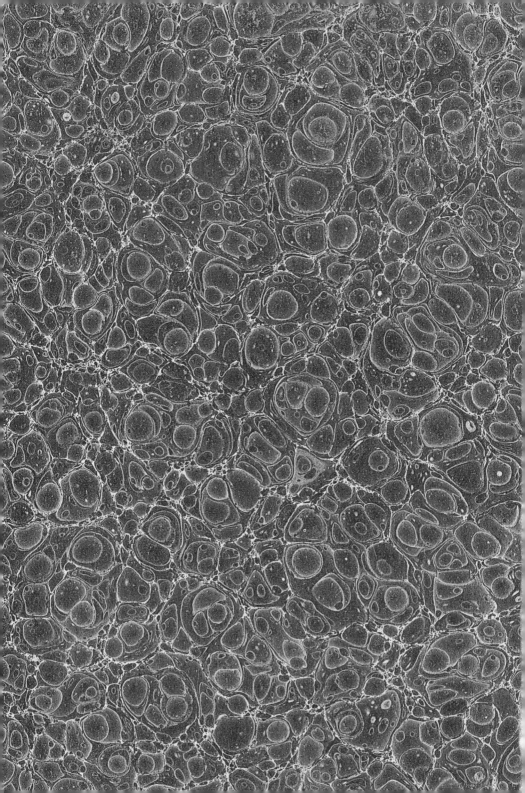

WRECK and SINKING
of the
TITANIC

WRECK and SINKING
of the
TITANIC

THE OCEAN'S
GREATEST DISASTER

A Graphic and Thrilling Account of the Sinking of the
Greatest Floating Palace Ever Built, Carrying Down to
Watery Graves More Than 1,500 Souls

Giving Exciting Escapes from Death and Acts of Heroism not
Equaled in Ancient or Modern Times, Told by

✟ THE SURVIVORS ✟

Including History of Icebergs, the Terror of the Seas,
Wireless Telegraphy and Modern Shipbuilding

Edited by
MARSHALL EVERETT
The Great Descriptive Writer

**HARPER
DESIGN**
An Imprint of HarperCollinsPublishers

WILLARD LIBRARY, BATTLE CREEK, MI

UNMARKED
○○○○○○○○○○○
SEPULCHRES

 h, what a burial was here! Not as when one is borne from his home among weeping throngs and gently carried to the green fields, and laid peacefully beneath the turf & flowers. No priest stood to perform a burial service. It was an ocean grave. The mists alone shrouded the burial place. No spade prepared the grave. No sexton filled up the hallowed earth. Down, down they sank, and the quick returning waters smoothed out every ripple and left the sea as though it had not been!

⊹

WILLARD LIBRARY, BATTLE CREEK, MI

DEDICATED
o o o o o o o o o o o

 o those who by their arts and deeds followed in the footsteps of HIM who suffered on the Cross, and who now sleep in unmarked sepulchres of the sea. Greater love hath no man, than this, that a man lay down his life for his friend.

—*St. John, 15 Chap., 13 V.*

CONTENTS

The Guardian Angel of the sea pays
tribute to the martyred heroes.

PREFACE

he disastrous collision, with an iceberg in mid-ocean, of the mighty ocean liner—the *Titanic*—the finest example of modern shipbuilding—and the awful loss of 1,595 of the 2,340 passengers aboard, goes down easily in history as the greatest of ocean catastrophes.

The *Titanic*, a floating palace of luxury, the largest and finest steamship ever built, set forth on her maiden trip with the avowed purpose of following the shortest course at the highest possible speed. Her 2,340 passengers, for the most part on pleasure bent, among whom were some of the wealthiest and most distinguished people of both sides of the Atlantic, felt such perfect confidence in the liner that, even after she was struck and when she was sinking, they could not believe her destruction possible.

But the "largest and finest steamship ever built," as she ploughed her way swiftly through the quiet Atlantic, plunged against a monster iceberg which lay stretched out for miles over the sea. Her side was ripped open, her boilers exposed to the icy waters, yet the people were not alarmed.

Not until the twenty lifeboats were put off with their protesting cargoes and the decks were tilted up dangerously did the people really lose their faith. The ship went down filling the quiet night with the cries of horror of the victims plunged to their untimely death.

Four agonizing hours were spent by the survivors suffering unbelievably from exposure and grief until they were picked up by the rescue ship *Carpathia*, utterly exhausted by their terrible experience. Thrilling tales of bravery and of sacrifice make this one of the most impressive tragedies of modern history. And the fact that the awful loss of life was avoidable by the simple provision of sufficient lifeboats to keep the passengers afloat till help could come, lays heavy indictment upon the modern commercial spirit which willingly takes in its care hundreds of people trusting to sheer luck to ward off danger.

This is the grim lesson to be learned from the tragedy. For the criminal neglect of others, these hundreds of innocent people atoned, and as an everlasting memorial to them will be the stringent laws made and enforced for safeguarding the millions who will trust their lives to the ocean in the future.

In this book the thrilling story is set forth clearly, the facts about the ship and the voyage, the passengers, the pathetic details of the wreck, the first-hand accounts of the survivors. The whole sad story is here.

CHAPTER I

THE TWO TITANS

s the *Titanic* drew away from the wharf to begin her only voyage, a common emotion quickened the thousands who were aboard her. Grimy slaves who worked and withered deep down in the glaring heat of her boiler rooms, on her breezy decks men of achievement and fame and millionaire pleasure seekers for whom the boat provided countless luxuries, in the steerage hordes of emigrants huddled in straited quarters but with their hearts fired for the new free land of hope; these, and also he whose anxious office placed him high above all—charged with the keeping of all of their lives—this care-furrowed captain on the bridge, his many-varied passengers, and even the remotest menial of his crew, experienced alike a glow of triumph born of pride in the enormous, wonderful new ship that carried them.

For she was the biggest boat that ever had been in the world. She implied the utmost stretch of construction, the furthest achievement in efficiency, the bewildering embodiment of an immense multitude of luxuries for which only the richest of the earth could pay. The cost of the *Titanic* was tremendous—it had taken many millions of dollars—many months to complete her. Besides (and best of all), she was practically unsinkable her owners said; pierce her hull anywhere, and behind was a watertight bulkhead, a sure defense to flout the floods and hold the angry ocean from its prey.

Angry is the word—for in all her triumph of perfection the *Titanic* was but man's latest insolence to the sea. Every article in her was a sheer defiance to the Deep's might and majesty.

The ship is not the ocean's bride; steel hull and mast, whirling shaft, and throbbing engine-heart (products, all, of serviceable wonderworking fire)—what kinship have these with the wild and watery waste? They are an affront and not an affinity for the cold and alien and elusive element that at all times threatens to overwhelm them.

But no one on the *Titanic* dreamed of danger when her prow was first set westward and her blades began the rhythmic beat that must not cease until the Atlantic had been crossed. Of all the statesmen, journalists, authors, famous financiers who were among her passengers (many of whom had arranged their affairs especially to secure passage in this splendid vessel), in all that brilliant company it may be doubted if a single mind secreted the faintest lurking premonition of a fear. Other ships could come safely and safely go, much more this monster— why, if an accident occurred and worse came to worst, she was literally too big to sink! Such was the instinctive reasoning of her passengers and crew, and such the unconsidered opinion of the world that read of her departure on the fatal day which marked the beginning of her first voyage and her last.

No doubt her very name tempted this opinion: *Titanic* was she titled—as though she were allied to the fabled giants of old called Titans, who waged war with the very forces of creation.

Out she bore, this giant of the ships, then, blithely to meet and buffet back the surge, the shock, of ocean's elemental might; latest enginery devised in man's eternal warfare against nature, product of a thousand minds, bearer of myriad hopes. And to that unconsidered opinion of the world she doubtless seemed even arrogant in her plenitude of power, like the elements she clove and rode—the sweeping winds above, the surging tide below. But this would be only in daytime, when the *Titanic* was beheld near land, whereon are multitudes of things beside which this biggest of the ships loomed large.

The noblest way for man to die
is when he dies for man.

Greater love hath no man.

When we imagine her alone, eclipsed by the solitude and immensity of night, a gleaming speck—no more—upon the gulf and middle of the vasty deep, while her gayer guests are dancing and the rest are moved to mirth or wrapped in slumber or lulled in security—when we think of her thus in her true relation, she seems not arrogant of power at all; only a slim and alien shape too feeble for her freight of precious souls, plowing a tiny track across the void, set about with silent forces of destruction compared to which she is as fragile as a cockle shell.

Against her had been set in motion a mass for a long time mounting, a century's stored-up aggregation of force, greater than any man-made thing as is infinity to one. It had expanded in the patience of great solitudes. On a Greenland summit, ages ago, avalanches of ice and snow collided, welded and then moved, inches in a year, an evolution that had naught to do with time. It was the true inevitable, gouging out a valley for its course, shouldering the precipices from its path. Finally the glacier reached the open Arctic, when a mile-in-width of it broke off and floated swinging free at last.

Does Providence directly govern everything that is? And did the Power who preordained the utmost second of each planet's journey, rouse up the mountain from its sleep of snow and send it down to drift, deliberately direct, into the exact moment in the sea of time, into the exact station in the sea of waters, where danced a gleaming speck—the tiny *Titanic*—to be touched and overborne?

It is easy thus to ascribe to the Infinite the direction of the spectacular phenomena of nature; our laws denote them "acts of God"; our instincts (after centuries of civilization) still see in the earthquake an especial instance of His power, and in the flood the evidence of His wrath. The floating menace of the sea and ice is in a class with these. The terror-stricken who from their ship beheld the overwhelming monster say that it was beyond all imagination vast and awful, hundreds of feet high, leagues in extent, black as it moved beneath no moon, appallingly suggestive of man's futility amidst the immensity of creation. See how, by a mere touch—scarcely a jar—one of humanity's proudest handiworks, the greatest vessel of all time, is cut down in her course, ripped up, dismantled and engulfed. The true Titan has overturned the toy.

Oh, where is now the boasted strength of that great hull of steel! Pitted against the iceberg's adamant it crumples and collapses. What of the ship unsinkable; assured so by a perfected new device? settling in the sea, shuddering to an inrush and an outburst of frigid water and exploding steam! All the effort of the thousand busy brains that built her, all the myriad hopes she bore—down, quite down! A long farewell to the toy Titan as the erasing waters fill and flatten smooth again to ocean's cold obliterating calm the handsbreadth she once fretted and defied!

Yes, it is easy to see God only in the grander manifestations of nature; but occasionally we are stricken by His speaking in the still small voice. Hundreds on this night of wreck were thus impressed. As the great steel-strong leviathan sank into the sea, those in the fleeing lifeboats heard, amidst the thunder and the discord of the monster's breaking-up, afar across the waters floating clear, a tremulous insistence of sweet sound, a hymn of faith—utterly triumphant o'er the solitudes! Men had left their work to perish and turned themselves to God.

When he builds and boasts of his *Titanic*, man may be great, but it is only when he is stripped of every cloying attribute of the world's pomp and power that he can touch sublimity. Those on the wreck had mounted to it from the time the awful impact came. The rise began when men of intellect and noted works, of titled place and honored station, worked as true yoke-fellows with the steerage passengers to see that all the women and their little ones were safely placed within the boats. They did this calmly, while the steamer settled low and every instant brought the waters nearer to their breath; exulting as each o'erburdened lifeboat safely drew away, and cheering until the iceberg echoed back the sound. There was very little fear displayed; calm intrepidity was here the mark of a high calling. Captain Smith, indeed, was afraid, but it was only for the precious beings under God committed to his care. And how manfully he minimized at first the danger until the rising surges creeping o'er the decks betrayed the awful truth. Then was the panic time! What cries were heard! What partings had and fond farewells! What love was lavished in renunciation and in life-and-death constancy when husband and wife refused to be separated in the hour that meant the inevitable death of one. But through all the time of terror the heroes of the *Titanic* remained true, nor yielded hearts to fear; and then, when all was done, when the last well-laden boat had safely put away, when the chill waters could be felt encroaching in the darkness, those who voluntarily awaited death, who had exemplified the sacred words: "Greater love hath no man than this, that a man lay down his life for a friend"—then these put heroism behind them for humility, rose to the greater height, threw themselves on Him who walked the waters to a sinking ship, as they sang in ecstasy the simple hymn of steadfast faith: "Nearer, my God, to Thee, nearer to Thee"!

Thus did man assert once more his high superiority among created things—he alone has power to revert to the unseen author of them all. Though compassed about with vast unfriendly Titans of the elements, builder himself of Toy Titans, like the boasted ship, that exist at the mercy of the sea and sky—at every fresh disaster that brings to nothingness his chiefest works, his spirit yet allies itself peculiarly with the power that only may be imagined and not seen; being persuaded that neither death, nor life, nor angels, nor principalities, nor powers, nor things present, nor things to come, nor height, nor depth, nor any other creature, shall be able to separate him from the love of God.

— *Fred S. Miller*

CHAPTER II

STORY OF THE TITANIC

THE "UNSINKABLE" TITANIC STRIKES AN ICEBERG AND SINKS
— HUNDREDS CARRIED TO SUDDEN AND UNTIMELY DEATH
AND FOR LACK OF ADEQUATE LIFESAVING SERVICE
— THE FACTS OF THE WRECK

he mighty ship *Titanic*, the triumph of the shipbuilders, thronged with happy, confident people, interested in her first voyage and her speed record, ploughed her swift way across the Atlantic, which lay smooth and calm and clear. In the midst of pleasant amusements and happy dreams there came a slight shock, a glancing blow from an iceberg, a few minutes of calm disbelief—then horror incredible. The Titan of nature and the *Titanic* of mechanical construction had met in mid-ocean. The iceberg ripped open the ship's side, exposing her boilers to the icy water, causing their explosion, plunging hundreds of people to their death within the short space of two hours. This is the tragic story of the beautiful ocean palace that sailed forth so gallantly from harbor on her maiden trip, April 10, 1912—buried under 2,000 fathoms of water with some 1,595 of her ill-fated passengers.

No more thrilling or pitiful tale has ever been written on the page of history—no greater record of human sacrifice and heroism.

The *Titanic* was the last word in shipbuilding and she set forth on her first voyage, the pride of an admiring world. Her luxurious appointments were beyond criticism, beautiful salons, reading and lounging rooms, palm courts, Turkish baths, private baths, a gymnasium, a swimming pool, a ballroom and billiard hall, everything one could imagine as making for comfort. Her mechanical construction was thought to be as perfect, and in the minds of her passengers was a faith in her "unsinkable" character almost unshakable. She carried nearly a full passenger roll, 2,340 people including the crew, as generally estimated, and was provided with only twenty lifeboats, sixteen ordinary lifeboats and four collapsible boats—accommodation for about a third of her passengers. These numbered some of the wealthiest and most prominent people on both sides of the Atlantic, John Jacob Astor, Major Archibald Butt, Benjamin Guggenheim, Isidor Straus, Charles M. Hays, Arthur Ryerson, Henry B. Harris, William T. Stead, Jacques Futrelle, and many more who gave up their lives in common with the humblest passengers in the steerage.

After the usual concert, Sunday evening, April 14, the passengers were in the midst of retiring or were amusing themselves in card and reading rooms. Some few were on deck enjoying the splendid evening, clear and fair, the ocean wonderfully calm. Suddenly there came a slight rocking of the ship, so slight as to be unnoticed by many. "Grazed an iceberg. Nothing serious," was the general comment as men resumed their interrupted card games. That was 11:40 P.M. Many people went to bed without another thought. The berg had been sighted only a quarter of a mile away, too late to check the ship's speed, so she rushed into the mass of ice, projecting only about eighty feet above sea level but reaching dangerously into the depths. The shock of the blow was so slight as to be scarcely perceptible to the unconscious passengers. But nevertheless it was a stroke dealing out death. For the *Titanic*, pushed on by her tremendous momentum of 21 knots an hour, sliding against the knife-like ledge, projecting unseen into the water, ripped her side open on the ice, shattering her airtight bulkheads. This permitted her gradual sinking, thereby allowing the icy waters to penetrate to her boilers, which had been working at high pressure, and causing their explosion, sending her to the bottom within two and one-half hours from the time she struck the iceberg.

Captain Smith took command as soon as the ship struck and the engines were stopped instantly. This sudden cessation of the constant vibration drew the passengers' attention more than did the shock of the collision. Life belts were ordered on the people immediately, and the boats were made ready, though the passengers thought all the time it was merely done for the sake of extraordinary precaution.

In the first boat the occupants were nearly all men, for there were no women on deck. The stewards and stewardesses were ordered below to summon the people from their staterooms, and when they came rushing out, some in their night clothes, some in evening gowns, all startled at the order even yet believing in the strength of the *Titanic*, the rule "women first" was rigidly enforced. Unwillingly the women were torn from their husbands, or placed in the boats by their husbands with the assurance that they would follow in other boats. In this way the boats were loaded with women and children, protesting but passive for the most part, with just two or three men to manage the oars. The scene was one of remarkable order. There was no mad struggle for safety; the men stood back and sent the women out, with very little disturbance. The report was circulated that the men and women were to be put in separate boats; also that there were boats on the other side of the ship and they were simply going later. Many thought, too, that their boats would soon be called back—that it was a mere matter of a short side-excursion. So the boasts were lowered away, and only when they were out in the water did their occupants realize the real danger. Then they could see the desperate plight of the *Titanic*.

As the *Titanic* sank gradually the water reached her engines, and an explosion tilted her decks, the list becoming more pronounced and consequently more dangerous every moment.

The sad partings—the last goodbyes.

Still the boats were loaded with women and children, until the last one swung off just in time.

The doomed multitude remaining shared her fate. Some leaped into the sea and clutched at floating wreckage; some sank with her, swimming to bits of wreckage as they struck the water; most of these were drowned, though a few escaped miraculously, picked up by the lifeboats or keeping themselves afloat by means of drifting boards and ship furnishings.

As the ship went down at 2:20 Monday morning, her colors flying, her captain in his place on the bridge, her bulk aglow with twinkling lights, the majority of her passengers looking out to sea from her decks, her string band playing "Nearer, My God, to Thee," united for the final moment the souls of the unhappy ones in safety of the frail boats with those loved ones helplessly going to their death. Then the lights winked, the black mass surged under and the death cries of the hundreds broke into the quite night. That was soon over, but the suffering in the lifeboats continued for hours. It was bitterly cold, due to the proximity of the iceberg; many of the boats were dashed partly full of the icy water; none of their occupants were sufficiently clad. In some of the boats, the women had to take the oars and they rowed with bleeding hands, these delicately nurtured ladies who proved their claim on heroism equal to that of the gentlemen. The boats were not provided with food, water, lighting facilities, necessities of any kind, and when the *Carpathia*, summoned by wireless, reached them, they could only signal by means of fragmentary letters and matches found about the persons of some of the passengers.

For four long hours they floated about, dazed by sorrow, nearly insensible from the bitter exposure to cold and wet, until the good ship *Carpathia* picked them up. Once in her cabins, they were given food and clothes; warmed, but not comforted. After the rescue, a service of thanksgiving, funeral service for the lost, was held—one of the most heart-breaking scenes ever enacted.

Thus ended the career of the *Titanic*, but her story will live long in the hearts of the survivors, and, to all the world, it bears a message that cannot be ignored—the message that to the god of commercial greed human sacrifices shall not be allowed at sea.

When the gallant ship *Titanic*, fair and false, set forth her initial trip with her 2,340 passengers, they little dreamed they were destined to point a moral to the world—that they were to be the instruments to demonstrate the criminal negligence of shipbuilders in deliberately sending forth vessels luxuriously equipped with every convenience and comfort, except the most essential one—lifeboats.

This great ocean liner—representing the acme of ship construction—went to her ruin after striking a huge iceberg in her course, an accident which probably was unavoidable, though greater care might have been exercised in the matter of speed.

To the twenty frail lifeboats fell the burden of keeping her 2,340 passengers afloat until the inevitable help should come, with the equally inevitable result that only 745 people emerged from the ill-fated wreck.

The cause for the disaster is undeniable; the reason for the loss of life is equally clear. The tales of horror of the survivors point to one single ominous fact; lack of adequate, commonsense protection of life paid to the Atlantic sea bottom the horrid toll of 1,595 persons.

Unequaled in their terrible, thrilling quality, the stories of this disaster; the striking of the iceberg, the loading of the boats, the agonized farewell, the mad leaps into the sea, the fearful hours upon the water before rescue and the bitter revelations of those lost, all these things stir the heart to sympathy and the conscience to a demand for lawful, law enforced safeguards that shall prevent another such grim tragedy.

These murdered hundreds were merely another instance of the innocent sacrifices offered to the god of commercial profit.

Someday, it is written, we shall cease this heathen worship; we shall demand proper precautions for our people, even though it be at the expense of the few paltry dollars. The time is now.

Laws shall be made and laws shall be enforced, and the future millions shall go to sea in ships provided with adequate safeguards. This is the service performed for us by these martyrs of the *Titanic*.

CHAPTER III

SPUR OF ICEBERG RIPPED OPEN BOTTOM OF THE TITANIC

GIGANTIC VESSEL LITERALLY DISEMBOWELED BY SUBMERGED FLOE WHILE SPEEDING — LITTLE SHOCK WAS FELT — PASSENGERS FOR HALF AN HOUR BELIEVED DAMAGE WAS SLIGHT AND TOOK THINGS CALMLY — MANY WERE IN IN THEIR STATEROOMS

It was the submerged spur of an iceberg of ordinary proportions that sent the White Star liner *Titanic* more than two miles to the bottom of the Atlantic off the banks of Newfoundland. The vessel was steaming almost full tilt through a gently swelling sea and under a starlit sky, in charge of First Officer Murdock, who a moment after the collision surrendered the command to Captain Smith, who went down with his boat.

The lifeboats that were launched were not filled to their capacity. The general feeling aboard the ship was, even after the boats had left its sides, that the vessel would survive its wound, and the passengers who were left aboard believed almost up to the last moment that they had a chance for their lives.

The captain and officers behaved with the utmost gallantry, and there was perfect order and discipline in the launching of the boats, even after all hope had been abandoned for the salvation of the ship and of those who were on board.

❈⊰ PLACID SEA HID DEATH ⊱❈

The great liner was plunging through a comparatively placid sea on the surface of which there was much mushy ice and here and there a number of comparatively harmless looking floes. The night was clear and stars visible. Chief Officer Murdock was in charge of the bridge.

The first intimation of the presence of the iceberg that he received was from the lookout in the crow's nest. They were so close upon the berg at this moment that it was practically impossible to avoid a collision with it.

The first officer did what other unstartled and alert commanders would have done under similar circumstances—that is, he made an effort by going full speed ahead on his starboard propeller and reversing his port propeller, simultaneously throwing his helm over, to make a rapid turn and clear the berg.

❈⊰ RIPPED BOTTOM OPEN ⊱❈

These maneuvers were not successful. He succeeded in preventing his bow from crashing into the ice cliff, but nearly the entire length of the great ship on the starboard side was ripped.

The speed of the *Titanic*, estimated to be at least twenty-one knots, was so terrific that the knifelike edge of the iceberg's spur protruding under the sea cut through her like a can opener.

The shock was almost imperceptible. The first officer did not apparently realize that the ship had received its death wound and none of the passengers it is believed had the slightest suspicion that anything more than a usual minor accident had happened. Hundreds who had gone to their berths and were asleep were not awakened by the vibration.

The destruction of the *Titanic*. The meeting of the Titans.

⫷ RETURNED TO CARD GAME ⫸

To illustrate the placidity with which practically all the men regarded the accident it was related that four who were in the smoking room playing bridge calmly got up from the table, and, after walking on deck and looking over the rail, returned to their game. One of them had left his cigar on the card table, and while the three others were gazing out on the sea he remarked that he couldn't afford to lose his smoke, returned for his cigar, and came out again.

The three remained only for a few moments on deck. They resumed their game under the impression that the ship had stopped for reasons best known to the commander and not involving any danger to her. The tendency of the whole ship's company except the men in the engine department, who were made aware of the danger by the inrushing water, was to make light of it and in some instances even to ridicule the thought of danger to so substantial a fabric.

⫷ SLOW TO REALIZE THE PERIL ⫸

Within a few minutes stewards and other members of the crew were sent around to arouse the people. Some utterly refused to get up. The stewards had almost to force the doors of the staterooms to make the somnolent appreciate their peril.

Mr. and Mrs. Astor were in their room and saw the ice vision flash by. They had not appreciably felt the gentle shock and supposed then nothing out of the ordinary had happened. They were both dressed and came on deck leisurely.

It was not until the ship began to take a heavy list to starboard that a tremor of fear pervaded it.

❧ LAUNCHED BOATS SAFELY ❧

The crew had been called clear away the lifeboats, of which there were twenty, four of which were collapsible. The boats that were lowered on the port side of the ship touched the water without capsizing. Some of the others lowered to starboard, including one collapsible, were capsized. All hands on the collapsible boats that practically went to pieces were rescued by other boats.

Sixteen boats in all got away safely. It was even then the general impression that the ship was all right and there is no doubt that that was the belief of even some of the officers.

At the lowering of the boats the officers superintending it were armed with revolvers, but there was no necessity for using them as there was nothing in the nature of a panic and no man made an effort to get into a boat while the women and children were being put aboard.

❧ BEGAN TO JUMP INTO SEA ❧

As the ship began to settle to starboard, heeling at an angle of nearly forty-five degrees, those who had believed it was all right to stick by the ship began to have doubt and a few jumped into the sea. These were followed immediately by others and in a few minutes there were scores swimming around. Nearly all of them wore life preservers.

One man who had a Pomeranian dog leaped overboard with it and striking a piece of wreckage was badly stunned. He recovered after a few minutes and swam toward one of the lifeboats and was taken aboard. Most of the men who were aboard the *Carpathia*, barring the members of the crew who had manned the boats, had jumped into the sea as the *Titanic* was settling.

Under instructions from officers and men in charge, the lifeboats were rowed a considerable distance from the ship itself in order to get away from the possible suction that would follow the foundering. The marvelous thing about the disappearance was so little suction as to be hardly appreciable from the point where the boats were floating. There was ample time to launch all boats before the *Titanic* went down, as it was two hours and twenty minutes afloat.

So confident were all hands that it had not sustained a mortal wound that it was not until 12:15 A.M., or thirty-five minutes after the berg was encountered, that the boats were lowered. Hundreds of the crew and a large majority of the officers, including Captain Smith, stuck to the ship to the last.

It was evident after there were several explosions, which doubtless were the boilers blowing up, that it had but a few minutes more of life.

❦◄ SINKS WITH LITTLE FLURRY ►❦

The sinking ship made much less commotion than the horrified watchers in the lifeboats had expected. They were close enough to the broken vessel to see clearly the most gruesome details of the foundering. All the spectators agreed that the shattered sections of the ship went down so quietly as to excite wonder.

Some of the rescued were scantily clad and suffered exceedingly from the cold, but the majority of them were prepared for the emergency. In the darkness aboard the ship that came shortly after the collision it was impossible for those in the boats to distinguish the identity of any of the persons who leaped into the sea. It is believed that nearly all cabin passengers who had not gone overboard immediately after the boats were launched vanished with the officers and crew.

❦⟨ HAD TIME TO DRESS ⟩❧

Some of the stewards who formed part of the lifeboat crew say that after the ship hit the berg the majority of the cabin passengers went back to their staterooms and that it was necessary to rout them out and in some instances force life preservers upon them. All agree that the engines of the ship were stopped immediately after she had made the ineffectual turn to clear the berg.

The lifeboats' crews were made up of stewards, stokers, coal trimmers, and ordinary seamen. It is said that the davits were equipped with a new contrivance for the swift launching of the boats, but that the machinery was so complicated and the men so unfamiliar with it that they had trouble in managing it.

CHAPTER IV

THRILLING STORY OF THE WRECK

TOLD BY L. BEASLEY, M. A., OF CAMBRIDGE UNIVERSITY, ENGLAND

he voyage from Queenstown had been quite uneventful; very fine weather was experienced and the sea was quite calm. The wind had been westerly to southwesterly the whole way, but very cold, particularly the last day; in fact, after dinner on Sunday evening it was almost too cold to be out on deck at all.

"I had been in my berth for about ten minutes when at about 11:40 P.M. I felt a slight jar and then soon after a second one, but not sufficiently large to cause any anxiety to anyone however nervous they may have been. The engines stopped immediately afterward and my first thought was—'she has lost a propeller.'

"I went up on the top deck in a dressing gown, and found only a few people there, who had come up similarly to inquire why we had stopped, but there was no sort of anxiety in the minds of anyone.

"We saw through the smoking-room window a game of cards going on and went in to inquire if the players knew anything; it seems they felt more of the jar, and looking through the window had seen a huge iceberg go by close to the side of the boat. They thought we had just grazed it with a glancing blow, and the engines had been stopped to see if any damage had been done. No one, of course, had any conception that she had been pierced below by part of the submerged iceberg.

"The game went on without any thought of disaster, and I retired to my cabin to read until we went on again. I never saw any of the players or the onlookers again. A little later, hearing people going upstairs, I went out again and found every one wanting to know why the engines had stopped.

"No doubt many were awakened from sleep by the sudden stopping of a vibration to which they had become accustomed during the four days we had been on board. Naturally, with such powerful engines as the *Titanic* carried, the vibration was very noticeable all the time, and the sudden stopping had something of the same effect as the stopping of a loud-ticking grandfather's clock in a room.

❦ Put on Life Belts ❦

"On going on deck again I saw that there was an undoubted list downward from stern to bow, but knowing of what had happened concluded some of the front compartments had filled and weighed her down. I went down again to put on warmer clothing, and as I dressed heard an order shouted:

"'All passengers on deck with life belts on.'

"We walked slowly up with them tied on over our clothing, but even then presumed this was a wise precaution the captain was taking, and that we should return in a short time and retire to bed.

"There was a total absence of any panic or any expressions of alarm, and I suppose this can be accounted for by the exceedingly calm night and the absence of any signs of the accident.

❖ Real Peril Was Hidden ❖

"The ship was absolutely still and except for a gentle tilt downward, which I do not think one person in ten would have noticed at that time, no signs of the approaching disaster were visible. She lay just as if she were waiting the other order to go on again when some trifling matter had been adjusted. But in a few moments we saw the covers lifted from the boats and the crews allotted to them standing by and curling up the ropes which were to lower them by the pulley blocks into the water.

"We then began to realize it was more serious than had been supposed, and my first thought was to go down and get more clothing and some money, but seeing people pouring up the stairs decided it was better to cause no confusion to people coming up by doing so. Presently we heard the order:

"'All men stand back away from the boats and all ladies retire to next deck below'—the smoking-room deck or B deck. The men all stood away and remained in absolute silence, leaning against the end railings of the deck or pacing slowly up and down.

"The boats were swung out and lowered from A deck. When they were to the level of B deck, where all the ladies were collected, the ladies got in quietly, with the exception of some who refused to leave their husbands. In some cases they were torn from them and pushed into the boats, but in many instances they were allowed to remain because there was no one to insist they should go.

"Looking over the side, one saw boats from aft already in the water, slipping quietly away into the darkness, and presently the boats near to me were lowered and with much creaking as the new ropes slipped through the pulley blocks down the ninety feet which separated them from the water. An officer in uniform came up as one boat went down and shouted: 'When you are afloat, row round to the companion ladder and stand by with the other boats for orders.'

❦ Discipline Holds Good ❧

"'Aye, aye, sir,' came up the reply, but I do not think any boat was able to obey the order. When they were afloat and had the oars at work the condition of the rapidly settling boat was so much more a sight for alarm for those in the boats than those on board that in common prudence the sailors saw they could do nothing but row from the sinking ship to save at any rate some lives. They no doubt anticipated that suction from such an enormous vessel would be more than usually dangerous to a crowded boat mostly filled with women.

"All this time there was no trace of any disorder, panic or rush to the boats, and no scenes of women sobbing hysterically, such as one generally pictures as happening at such times; every one seemed to realize so slowly that there was imminent danger.

"When it was realized that we might all be presently in the sea, with nothing but our life belts to support us until we were picked up by passing steamers, it was extraordinary how calm every one was and how completely self-controlled.

"One by one the boats were filled with women

and children, lowered and rowed away into the night. Presently the word went round among the men, 'the men are to be put into the boats on the starboard side.' I was on the port side, and most of the men walked across the deck to see if this was so.

"I remained where I was, and presently heard the call:

"'Any more ladies?' Looking over the side of the ship, I saw the boat, No. 13, swinging level with B deck, half full of ladies.

"Again the call was repeated:

"'Any more ladies?'

"I saw none come on and then one of the crew looked up and said: 'Any ladies on your deck, sir?'

"'No,' I replied.

"'Then you had better jump.'

"I dropped in and fell in the bottom, as they cried 'lower away.' As the boat began to descend two ladies were pushed hurriedly through the crowd on B deck and heaved over into the boat, and a baby of ten months passed down after them. Down we went, the crew calling to those lowering which end to keep her level. 'Aft,' 'stern,' 'both together,' until we were some ten feet from the water, and here occurred the only anxious moment we had during the whole of our experience from leaving the deck to reaching the *Carpathia*.

❧ New Peril Threatened ☙

"Immediately below our boat was the exhaust of the condensers, a huge stream of water pouring all the time from the ship's side just above the water line. It was plain we ought to be quite a way from this not to be swamped by it when we touched water. We had no officer aboard, nor petty officer or member of the crew to take charge. So one of the stokers shouted: 'Some one find the pin which releases the boat from the ropes and pull it up.' No one knew where it was. We felt as well as we could on the floor and sides, but found nothing, and it was hard to move among so many people—we had sixty or seventy on board.

"Down we went and presently floated with our ropes still holding us, the exhaust washing us away from the side of the vessel and the swell of the sea urging us back against the side again. The result of all these forces was an impetus, which carried us parallel to the ship's side and directly under boat No. 14, which had filled rapidly with men and was coming down on us in a way that threatened to submerge our boat.

❦ Sound Failed to Carry ❧

"'Stop lowering 14,' our crew shouted, and the crew of No. 14, now only twenty feet above, shouted the same. But the distance to the top was some seventy feet and the creaking pulleys must have deadened all the sound to those above, for down it came—fifteen feet, ten feet, five feet, and a stoker and I reached up and touched her swinging above our heads, but just before it dropped another stoker sprang to the ropes with his knife.

"'One,' I heard him say; 'two,' as his knife cut through the pulley ropes, and the next moment the exhaust steam had carried us clear, while boat 14 dropped into the space we had the moment before occupied, our gunwales almost touching. We drifted away easily as the oars were got out and headed directly

At dawn the lights of the rescuing ship, *Carpathia*, appeared on the horizon.

away from the ship. The crew seemed to me to be mostly cooks in white jackets, two to an oar, with a stoker at the tiller.

"The captain-stoker told us that he had been on the sea twenty-six years and had never seen such a calm night on the Atlantic. As we rowed away from the *Titanic* we looked back from time to time to watch it, and a more striking spectacle it was not possible for any one to see.

❖─ Titanic Great in Death ─❖

"In the distance it looked an enormous length, its great bulk outlined in black against the starry sky, every porthole and saloon blazing with light. It was impossible to think anything could be wrong with such a leviathan were it not for that ominous tilt downward in the bow, where the water was by now up to the lowest row of portholes. Presently about 2 A.M., as near as I can remember, we observed it settling very rapidly, with the bow and bridge completely under water, and concluded it was now only a question of minutes before it went; and so it proved.

"It slowly tilted straight on end, with the stern vertically upward, and as it did, the lights in the cabins and saloons, which had not flickered for a moment since we left, died out, came on again for a single flash, and finally went altogether.

"To our amazement the *Titanic* remained in the upright position, bow down, for a time which I estimate as five minutes, while we watched at least 150 feet of the *Titanic* towering above the level of the sea and looming black against the sky. Then the ship dived beneath the waters.

❦⟶ Heard Cry of Dying ⟶❦

"And then, with all these, there fell on the ear the most appalling noise that human being ever listened to—the cries of hundreds of our fellow beings struggling in the icy cold water, crying for help with a cry that we knew could not be answered. We longed to return and pick up some of those swimming, but this would have meant swamping our boat and loss of life to all of us.

❦⟶ The *Carpathia* Appears ⟶❦

"Our rescuer showed up in a few hours, and as it swung round we saw its cabins all alight and knew it must be a large steamer. It was now motionless, and we had to row to it. Just then day broke, a beautiful, quiet dawn with faint pink clouds just above the horizon, and a new moon whose crescent just touched the waters.

"The passengers, officers and crew gave up gladly their staterooms, clothing and comforts for our benefit, all honor to them."

The English Board of Trade passenger certificate on board the *Titanic* showed approximately 3,500. The same certificate called for lifeboat accommodation for approximately 950 in the following boats:

Fourteen large lifeboats, two smaller boats and four collapsible boats. Life preservers were accessible and apparently in sufficient numbers for all on board.

The approximate number of passengers carried at the time of collision was:

First class, 330; second class, 320; third class, 750; total, 1,400. Officers and crew, 940. Total, 2,340.

Of the foregoing about the following were rescued by the steamship *Carpathia*:

First class, 210; second class, 125; third class, 200; officers, 4; seamen, 39; stewards, 96; firemen, 71; total, 210 of the crew. The total, about 745 saved, was about 80 percent of the maximum capacity of the lifeboats.

CHAPTER V

RESCUE OF THE SURVIVORS

ONLY 745 OF THE 2,340 SOULS ABOARD DOOMED LINER SAVED BY THE LIFEBOATS — LITTLE SHOCK FELT WHEN THE ICEBERG WAS STRUCK BY THE TITANIC

Freighted with its argosy of woe, disaster and death, bringing glad reunion to some, but misery unutterable to many, the *Carpathia*, with the survivors of the lost *Titanic* aboard, came back to a grief-stricken city and nation four days after the disaster. It was received by awe-stricken thousands whose conversation was conducted in whispers.

The story it brought home was one to crush the heart with its pathos, but at the same time to thrill it with pride in the manly and womanly fortitude displayed in the face of the most awful peril and inevitable death.

As the *Titanic* went down, according to the story of those who were among the last to leave the wounded hulk, the ship's band was playing.

❦ ESTIMATED 1,595 DEAD ❦

As brought to port by the *Carpathia*, the death list was placed at 1,601. The *Titanic* had aboard 2,340 persons, of whom 745 were picked up. Six of the latter succumbed to the exposure they had undergone before the *Carpathia* reached port. Not only was the *Titanic* tearing through the April night to its doom with every ounce of steam crowded on, but it was under orders from the general officers of the line to make all the speed of which she was capable. This was the statement made by J. H. Moody, a quartermaster of the vessel and helmsman on the night of the disaster. He said the ship was making twenty-one knots an hour, and the officers were striving to live up to the orders to smash the record.

"It was close to midnight," said Moody, "and I was on the bridge with the second officer, who was in command. Suddenly he shouted, 'Port your helm!' I did so, but it was too late. We struck the submerged portion of the berg."

◈◄ LITTLE SHOCK FELT ►◈

As nearly as most of the passengers could remember, the *Titanic*, sliding through the water at no more speed than had been consistently maintained during all of the trip, slid gracefully a few feet out of the water with just the slightest tremble. It rolled slightly; then it pitched. The shock, scarcely noticeable to those on board, drew a few loungers over to the railings. Officers and petty officers were hurrying about. There was no destruction within the ship, at least not in the sight of the passengers.

There was no panic. Everything that could be seen tended to alleviate what little fear had crept into the minds of the passengers who were more apprehensive than the regular travelers who cross the ocean at this season of the year and who were more used to experiencing those small quivers.

Not one person aboard the *Titanic*, unless possibly it was the men of the crew, who were working far below, knew the extent of the injuries it had sustained. Many of the passengers had taken time to dress, so sure were they that there was no danger. They came on deck, looked the situation over and were unable to see the slightest sign that the *Titanic* had been torn open beneath the water line.

When the passengers' fear had been partly calmed and most of them had returned to their staterooms or to the card games in which they were engaged before the quiver was felt, there came surging through the first cabin quarters a report that seemed to have drifted in from nowhere that the ship was sinking. How this word crept in from outside no one seems now to know. Immediately the crew began to man the boats.

Then came the shudder of the riven hulk of the once magnificent steamship as it receded from the shelving ice upon which it had driven, and its bow settled deeply into the water.

"We're lost! We're lost!" was the cry that rose from hundreds of throats. "The ship is sinking. We must drown like rats!"

Women in evening gowns, with jewels about their necks, knelt on deck, amid the vast, fear-stricken throng, crowded about the lifeboats and prayed for help. Others, clad in their night clothing, begged the officers to let them enter the boats.

"Everybody to the boats!" was the startling cry that was repeated from end to end of the *Titanic*.

"Women and children first!" was the hoarse order that went along the line of lifeboats.

Without food, without clothing and with only the clothes in which they stood when the shock came, the women were tossed over the rails of the lifeboats, the davits were swung out, a few men were picked to man the oars, an officer to command the boat and the order to "lower away" was shouted. The little craft, laden with living freight, were launched.

❮ NO CHOICE BETWEEN CLASSES ❯

Men whose names and reputation were prominent in two hemispheres were shouldered out of the way by roughly dressed Slavs and Hungarians. Husbands were separated from their wives in the battle to reach the boats. Tearful leave-takings as the lifeboats, one after another, were filled with sobbing women and lowered upon the ice-covered surface of the ocean were heartbreaking.

There was no time to pick or choose. The first woman to step into a lifeboat held her place even though she was a maid or the wife of a Hungarian peasant. Many women clung to their husbands and refused to be separated. In some cases they dragged their husbands to the boats and in the confusion the men found places in the boats.

Before there was any indication of panic, Henry B. Harris, a theatrical manager of New York, stepped into a boat at the side of his wife before it was lowered.

"Women first!" shouted one of the ship's officers. Mr. Harris glanced up and saw that the remark was addressed to him.

"All right," he replied, coolly.

"Goodbye, my dear," he said, as he kissed his wife, pressed her a moment to his breast and then climbed back to the *Titanic*'s deck.

◄◄ FLEET DREW AWAY ►►

One by one the little fleet drew away from the towering sides of the giant steamship, whose decks were already reeling as it sank lower in the water.

"The *Titanic* is doomed!" was the verdict that passed from lip to lip.

"We will sink before help can come!"

Water poured into every compartment of the 800-foot hull, where great plates had been torn apart and huge rivets were sheared off as though they were so much cheese.

Pumps were started in the engine-room, but the water poured into the great hull in such torrents through scores of rents that all knew the fight to save the steamship was hopeless.

Overhead the wireless buzzed the news to the other steamships. The little fleet of lifeboats withdrew to a safe distance and the 1,595 left on board with no boats waited for the merciful death plunge, which ended all.

❦ WOMEN SAVED FIRST ❦

A few spars, a box or two, a few small pieces of other wreckage, were the only portions of the *Titanic* corpse that lived on the water surface to be beheld by the persons on board the *Carpathia* when it rushed to the rescue. It was just breaking day as the rescue work was completed.

So exhausted were the survivors that scarcely any of them were able to tell their story of what actually had happened until late in the afternoon of Monday. It seemed impossible to obtain a complete story of the tragedy.

Survivors in a lifeboat. Photograph taken from the *Carpathia*.

❊◄ FEW INJURED ON WRECK ►❊

Certainly few of the *Titanic* passengers were hurt on board that great vessel. Few of the persons who came in among the survivors on the *Carpathia* bore any marks of injury. Their sufferings were caused chiefly by exposure, shock and grief. The latter was terrible. Many of the women had walked into a boat after kissing their husbands goodbye.

The women in the lifeboats saw their loved ones plunge to death. The survivors' boats were bobbing along in the waves all within a radius of half a mile of the great *Titanic*, when, with a roar and burst of spray, it settled and passed out of sight for the last time.

Then began one of the most torturous experiences for the helpless women in the drifting lifeboats that human beings ever were compelled to endure.

It was black night. Fortunately several of the men who were saved and some of the few petty officers who had aided in manning the lifeboats had a few matches in their pockets. Their torches were improvised from letters and scraps of papers that were found in their pockets. There was nothing to be seen.

❊◄ SIGNALED WITH TORCHES ►❊

The torches, the only hope of those who thought they were doomed to death, were being carefully guarded and many times those who held them were implored to light them in the faint hope that rescue was closer at hand than even the most sanguine could have believed.

But the strong prevailed and it was not until the first rocket was seen to shoot heavenward from the *Carpathia* that the first of the torches was lighted and its filmy blaze shot up as high as was possible when one of the men, held on the shoulders of five others, stood up and waved the flaming papers until they burned down to his fingertips.

The desolate groups huddled together in the tossing and rolling tiny craft could not tell whether their torch had been seen by the ship that was firing the rockets. They waited fifteen minutes and the operation was repeated.

Then the huge bulk of the *Carpathia* took form in the gray of the breaking morning and it swept swiftly down into the center of a widely separated fleet of lifeboats with their human freight, then more dead than alive. They had been for approximately six hours in the open with the waves sending spray and at intervals whole barrelsful of water in upon them. They were drenched and the severe cold was freezing their clothing to their bodies. Only a few of them were able to walk when it finally came their turn to be taken on board the *Carpathia*.

The *Carpathia*'s sailors went after those lying unconscious in the bottom of the lifeboats, lifted them up to other sailors standing on the *Carpathia*'s ladders. Everything that could be done for the survivors was done on the *Carpathia*.

Several of them had been cut and bruised in their attempts to get into the lifeboats and by falling from exhaustion during the awful ordeal they were compelled to pass through while waiting for the *Carpathia* to come to their relief. These were given surgical care. The others were placed in bed and few if any of them were able during the rest of the voyage to go on deck.

❦ TELLS OF THE RESCUE ❧

A passenger on the *Carpathia* made the following statement:

"I was awakened at about half past twelve at night by a commotion on the decks which seemed unusual, but there was no excitement. As the boat was moving I paid little attention to it, and went to sleep again. About three o'clock I again awakened.

I noticed that the boat had stopped. I went to the deck. The *Carpathia* had changed its course.

Lifeboats were sighted and began to arrive—and soon, one by one, they drew up to our side. There were sixteen in all, and the transferring of the passengers was most pitiable. The adults were assisted in climbing the rope ladder by ropes adjusted to their waists. Little children and babies were hoisted to the deck in bags.

Survivors of the *Titanic* sit on board the *Carpathia*.

⫷ Few in Some Boats ⫸

"Some of the boats were crowded, a few were not half full. This I could not understand. Some people were in full evening dress. Others were in their nightclothes and were wrapped in blankets. These, with immigrants in all sorts of shapes, were hurried into the saloon indiscriminately for a hot breakfast. They had been in the open boats four and five hours in the most biting air I ever experienced.

"There were husbands without wives, wives without husbands, parents without children and children without parents. But there was no demonstration. No sobs—scarcely a word spoken. They seemed to be stunned. Immediately after breakfast, divine service was held in the saloon.

"One woman died in the lifeboat; three others died soon after reaching our deck. Their bodies were buried in the sea at five o'clock that afternoon. None of the rescued had any clothing except what they had on, and a relief committee was formed and our passengers contributed enough for their immediate needs.

❖ Tells of the Final Plunge ❖

"When its lifeboats pushed away from the *Titanic*, the steamer was brilliantly lighted, the band was playing and the captain was standing on the bridge giving directions. The bow was well submerged and the keel rose high above the water. The next moment everything disappeared. The survivors were so close to the sinking steamer that they feared the lifeboats would be drawn into the vortex.

"On our way back to New York we steamed along the edge of a field of ice which seemed limitless. As far as the eye could see to the north there was no blue water. At one time I counted thirteen icebergs."

CHAPTER VI

SURVIVORS REACH NEW YORK

HOSPITALS SENT AMBULANCES AND NURSES — INVESTIGATION BY THE SENATE DECIDED UPON

t 8 o'clock automobiles and carriages containing relatives and friends of the survivors began arriving at the White Star pier. When the *Carpathia* was sighted coming up the river at 8:45 more than 500 automobiles and other vehicles were packed within the police lines.

Significant of the tragic side of the event was the frequent arrivals of ambulances and auto trucks from all the big department stores, filled with cots, invalid chairs and surgical appliances. Right of way was given the ambulances and they were permitted to park directly alongside the pier entrance.

❦ HOSPITALS SENT NURSES ❧

From St. Vincent's Hospital came twelve black-robed sisters to nurse the injured, and all the ambulances of the institution except one. The full surgical staff of the hospital also was in attendance. Ambulances and surgeons were on hand from St. Luke's Hospital, Bellevue, Roosevelt and Flower hospitals, and a great number of physicians who had volunteered their services.

The Sisters of Charity found work to do before the arrival of the *Carpathia*. Women in the throng awaiting relatives became hysterical with dread and anxiety and the black-robed sisters went to them, put their arms about them and comforted them and administered restoratives.

Eva Booth, commander of the Salvation Army, and fifty assistants, who meet all incoming vessels to minister to immigrants, were allowed within the police lines, but they were turned back at the entrance of the Cunard pier and only Miss Booth and three of her party were admitted.

Curiosity seekers standing in front of the Cunard piers at 7 A.M. to await the *Carpathia* returning with survivors from the *Titanic*.

❦ BROKERS BROUGHT $20,000 ❦

Among those on the pier were six members of the New York Stock Exchange, with $20,000, which had been collected on the floor of the exchange. They had instructions to use the money among the steerage passengers in any way they saw fit.

The women of the relief committee to look after the steerage passengers arrived in autos and theater buses, in which the sufferers were to be taken to hospitals. Gimbel Brothers sent all their delivery wagons to the pier, laden with first aid appliances and cots, and placed them at the disposal of the women's relief committee. In addition, the firm announced they would provide quarters for two hundred sufferers overnight in their store.

◄◄ CALLED FOR MORE NURSES ►►

Relatives and friends of the survivors had reached the pier before half past eight o'clock, but for another half hour automobiles arrived containing physicians and nurses and loaded with first aid appliances. The surgeons and nurses were in working attire, the women in white gowns and caps, the surgeons in white duck trousers and jackets.

A party of four surgeons and ten nurses arrived in three automobile buses and as they hurried to the pier one of them said they had been sent by Mrs. William K. Vanderbilt.

In spite of the number of physicians that had reached the pier at 8:30, it was found there was a dearth of nurses and hurried calls were sent out to all the city institutions and private hospitals and nurses' exchanges. In response to these calls nurses began arriving in taxicabs and autos, and before the *Carpathia* was warped into its pier there were more than two hundred nurses awaiting to go on board.

Ropes dotted with green lights were stretched for seventy-five yards in front of the piers to hold back the throngs. No one without a special permit was allowed beyond these ropes.

The Pennsylvania Railroad Company had a special train waiting at its station at Thirty-fourth Street and a number of taxicabs to convey survivors desiring to go to Philadelphia to their friends.

News that the *Carpathia* was outside of the harbor and rapidly approaching sent thousands of persons to vantage points along the city's waterfront. At the Battery, the first point on Manhattan Island which the rescue ship would pass, a crowd estimated at ten thousand persons assembled. Other vantage points further uptown were crowded with spectators eager to catch the first glimpse of the approaching *Carpathia*.

❧ INVESTIGATION DECIDED ON ❧

Senator William Alden Smith of Michigan and Senator Newlands of Nevada arrived in New York at 9 P.M. April 18 to summon survivors of the *Titanic* and officials of the International Mercantile Marine to testify before the Senate subcommittee appointed to investigate the disaster of the sea.

When the senators arrived at the Pennsylvania station they were informed that the *Carpathia* was at its pier. The committee had intended boarding a revenue cutter and going down the bay to meet the *Carpathia* and boarding it. Upon learning this the senators secured cabs and announced they were going direct to the pier.

CHAPTER VII

LAST MAN OFF TELLS HORRORS OF SHIPWRECK

COLONEL GRACIE, U.S.A., RESCUED AFTER GOING DOWN
ON TITANIC'S TOPMOST DECK — HEROES ON ALL SIDES
— MRS. ISIDOR STRAUS DROWNED, REFUSING TO DESERT
HUSBAND — ASTOR PRAISED FOR CONDUCT

 olonel Archibald Gracie, U.S.A., the last man saved after the wreck of the *Titanic*, went down with the vessel, but was picked up. He was met in New York by his daughter and his son-in-law, Paul H. Fabricius.

Colonel Gracie told a remarkable story of personal hardship and denied emphatically reports that there was any panic on board the steamship after the disaster. He praised in the highest terms the behavior of both the passengers and crew and paid a high tribute to the heroism of the women passengers.

"Mrs. Isidor Straus," said Colonel Gracie, "went to her death because she would not desert her husband. Although he pleaded with her to take her place in the boat, she steadfastly refused, and when the ship settled at the head the two were engulfed by the wave that swept the vessel."

❖⊰ DRIVEN TO TOP DECK ⊱❖

Colonel Gracie told how he was driven to the top-most deck when the ship settled and was the sole survivor after the wave that swept it just before its final plunge had passed.

"I jumped with the wave just as I often have jumped with the breakers at the seashore. By great good fortune I managed to grasp the brass railing on the deck above, and I hung on by might and main.

"When the ship plunged down I was forced to let go and was swirled around and around for what seemed to be an interminable time. Eventually I came to the surface to find the sea a mass of tangled wreckage.

"Luckily, I was unhurt, and, casting about, managed to seize a wooden grating floating nearby. When I had recovered my breath I discovered a canvas and cork life raft which had floated up.

❖⊰ Thirty Saved on Raft ⊱❖

"A man whose name I did not learn was struggling toward this raft from some wreckage to which he had clung. I cast off and helped him get onto the raft, and we then began the work of rescuing those who had jumped into the sea and were floundering in the water.

"When dawn broke there were thirty of us on the raft, standing knee-deep in the icy water and afraid to move lest the craft be overturned.

"Several other unfortunates, benumbed and half dead, besought us to save them, and one or two made efforts to reach us, but we had to warn them away. Had we made any effort to save them, we all might have perished.

⟨⟨ Long Hours of Horror ⟩⟩

"The hours that elapsed before we were picked up by the *Carpathia* were the longest and most terrible that I ever spent. Practically without any sensation of feeling because of the icy water, we were almost dropping from fatigue.

"We were afraid to turn around to learn whether we were seen by passing craft, and when some one who was facing astern passed the word that something that looked like a steamer was coming up one of them became hysterical under the strain. The rest of us, too, were nearing the breaking point."

Colonel Gracie denied with emphasis that any men were fired upon, and declared that only once was a revolver discharged.

"This," the colonel said, "was done for the purpose of intimidating some steerage passengers who had tumbled into a boat before it was prepared for launching. The shot was fired in the air, and when the foreigners were told that the next would be directed at them they promptly returned to the deck. There was no confusion and no panic."

Contrary to the general expectation, there was no jarring impact when the vessel struck, according to the army officer. He was in his berth when the *Titanic* smashed into the submerged portion of the iceberg and was aroused by the jar.

⟨⟨ STOPPED WATCH FIXED TIME ⟩⟩

Colonel Gracie looked at his watch, he said, and found it was just midnight. The ship sank with him at 2:22 A.M., for his watch stopped at that hour. He described these events:

"Before I retired, I had a long chat with Charles M. Hays, president of the Grand Trunk Railroad. One of the last things Mr. Hays said was this:

"'The White Star, the Cunard and the Hamburg-American lines are devoting their attention and ingenuity to vying with one another to attain supremacy in luxurious ships and in making speed records. The time will soon come when this will be checked by some appalling disaster.'

"Poor fellow—a few hours later he was dead.

✦ Gave Praise to Astor ✦

"The conduct of Colonel John Jacob Astor was deserving of the highest praise," Colonel Gracie declared.

"The millionaire New Yorker devoted all his energies to saving his young bride, formerly Miss Force of New York, who was in delicate health.

"Colonel Astor helped us in our efforts to get Mrs. Astor in the boat. I lifted her into the boat and as she took her place Colonel Astor requested permission of the second officer to go with her for her own protection.

"'No, sir,' replied the officer, 'not a man shall go on a boat until the women are all off.'

✦ Colonel Aided with Boats ✦

"Colonel Astor then inquired the number of the boat which was being lowered away and turned to the work of clearing the other boats and reassuring the frightened and nervous women.

"By this time the ship had begun to list frightfully

to port. This became so dangerous that the second officer ordered every one to rush to starboard.

"This we did, and we found the crew trying to get a boat off in that quarter. Here I saw the last of John B. Thayer and George B. Widener of Philadelphia."

❖⎯ IGNORED WARNINGS CHARGED ⎯❖

Colonel Gracie said that despite the warnings of icebergs no slowing down of speed was ordered by the commander of the *Titanic*. There were other warnings, too, he said.

"In the twenty-four hours' run ending the 14th, the ship's run was 546 miles, and we were told that the next twenty-four hours would see even a better record posted.

"No diminution of speed was indicated in the run and the engines kept up their steady work. When Sunday evening came we all noticed the increased cold, which gave plain warning that the ship was in close proximity to icebergs or ice fields. The officers, I am credibly informed, had been advised by wireless from other ships of the presence of icebergs and dangerous floes in the vicinity. The sea was as smooth as glass and the weather clear, so that it seemed that there was no occasion for fear.

"When the vessel struck it, the passengers were so little alarmed that they joked over the matter. The few who appeared upon deck early had taken their time to dress properly and there was not the slightest indication of panic. Some fragments of ice had fallen on the deck and these were picked up and passed around by some of the facetious ones, who offered them as mementos of the occasion. On the port side, a glance

Colonel John Jacob Astor, who was lost
with the *Titanic*, and his wife, Madeline Force, who was rescued.

over the side failed to show any evidence of damage, and the vessel seemed to be on an even keel.

"James Clinch Smith and I, however, soon found the vessel was listing heavily. A few minutes later the officers ordered men and women to don life preservers."

❈❮ WOMEN REFUSED RESCUE ❯❈

One of the last women seen by Colonel Gracie, he said, was Miss Evans, of New York, who virtually refused to be rescued, because "she had been told by a fortune teller in London that she would meet her death on the water."

A young Englishwoman who requested that her name be omitted told a thrilling story of her experience in one of the collapsible boats, which was manned by eight of the crew from the *Titanic*. The boat was in command of the fifth officer, H. Lowe, whom she credited with saving the lives of many persons.

Before the lifeboat was launched Lowe passed along the port deck of the steamer, commanding the people not to jump into the boats and otherwise restraining them from swamping the craft. When the collapsible was launched Lowe succeeded in putting up a mast and a small sail.

The officer collected the other boats together, and, in case where some were short of adequate crews, directed an exchange by which each was adequately manned. He threw lines which linked the boats two by two, and all thus moved together.

Later on Lowe went back to the wreck with the crew of one of the boats and succeeded in picking up some of those who had jumped overboard, and were swimming about. On his way back to the *Carpathia* he passed one of the collapsible boats which was on the point of sinking with thirty passengers aboard, most of them in scant night clothing. They were rescued just in the nick of time.

CHAPTER VIII

HEROISM ON THE TITANIC

PRESIDENT TAFT'S ESTIMATE OF MAJOR BUTT — BEN
GUGGENHEIM NOT A COWARD — HEROIC MUSICIANS —
"NEARER, MY GOD, TO THEE."

hen President Taft heard that women and children had perished in the wreck of the *Titanic* he spoke his estimate of Archie Butt in saying: "I do not expect, I do not want, to see him back." That Mr. Taft knew his man was proved by the words of the rescued.

Note this: Benjamin Guggenheim sent word to his wife: "Tell her I played the game out straight to the end. No women shall be left aboard this ship because Ben Guggenheim was a coward."

And this: "And then Mrs. Straus would call him (Mr. Straus) by his first name and say her place was with him, that she had lived with him and that she would die with him." And Mr. Straus said: "I am not too old to sacrifice myself for a woman."

And this of Mrs. Allison: "The boat was full and she grasped Lorraine with one arm and her husband with the other and stood smiling as she saw us rowing away."

And this of Captain Smith: "He swam to where a baby was drowning, carried it in his arms to a lifeboat, and then swam back to his ship to die." And this, the command given by Captain Smith bringing order out of chaos: "Be British, my men."

And lastly: Kraus, Hume, Taylor, Woodward, Clark, Brailey, Breicoux and Hartley, when the last faint hope was gone, lined up on deck, stood in water up to their knees and played "Nearer, My God to Thee," as 1,500 souls passed from life.

◄— HEROIC MUSICIANS —►

Except in the case of the English ship *Birkenhead*, when the soldiers on board stood at parade after the women and children had been taken into the boats and the band played the national air as the ship went down, we do not recall a parallel to the conduct of the musicians on board the *Titanic*, who, as all accounts agree, ceased not their inspiriting ministrations until they were engulfed by the waves.

Indeed, it seems even to be a question if the later instance of heroism was not greater than the former, for the bandsmen on the *Birkenhead* were enlisted men, obeying orders like soldiers, while it is scarcely to be thought that the obligations of the musicians on the *Titanic* required them to play with death confronting them. There has been a marvelous upwelling of sympathy for the families made destitute by the awful catastrophe, and, perhaps, a too great multiplicity of relief funds; but there is, nevertheless, something especially appealing in Dr. Frank Damrosch's suggestion that a special contribution be asked for the families of those who gave courage and comfort to the doomed victims of the steamship; and died to do it.

◄— MAJOR BUTT DIED LIKE A SOLDIER —►

A graphic story of heroism of Major Archibald W. Butt on the *Titanic* was told feelingly by Miss Marie Young, a former resident of New York, before going to her home in Washington, D.C. Miss Young is believed to have been the last woman to leave the *Titanic* and the last of the survivors to have talked with the President's military aide. She and Major Butt had long been friends, Miss

Purser McElroy and Captain Smith, of the ill-fated *Titanic*.

Young having been a special music instructor to the children of Theodore Roosevelt when he was President. Miss Young said:

"The last person to whom I spoke on board the *Titanic* was Archie Butt, and his good, brave face smiling at me from the deck of the steamer was the last I could distinguish as the lifeboat I was in pulled away from the steamer's side.

"Archie himself put me into the boat, wrapped blankets around me and tucked me in as carefully as if we were starting on a motor ride. He himself entered the boat with me, performing the little courtesies as calmly and with as smiling a face as if death was far away instead of being but a few moments removed from him. When he had carefully wrapped me up he stepped on the gunwale of the boat, and, lifting his hat, smiled down at me.

"'Goodbye, Miss Young,' he said, bravely and smilingly. 'Luck is with you. Will you kindly remember me to all the folks back home?'

"Then he stepped to the deck of the steamer, and the boat I was in was lowered to the water. It was the last boat to leave the ship; of this I am perfectly certain. And I know that I am the last of those who were saved to whom Archie Butt spoke. As our boat was lowered and left the side of the steamer Archie was still standing at the rail looking down at me. His hat was raised, and the same old, genial, brave smile was on his face. The picture he made as he stood there, hat in hand, brave and smiling, was one that will always linger in my memory."

Mrs. Henry B. Harris, in an interview, also described the heroism of Major Butt. She said:

"Archie Butt was a major to the last. God never made a finer nobleman than he. The sight of that man, calm, gentle, and yet as firm as a rock, never will leave me. The American Army is honored by him, and the way he showed some of the other men how to behave when women and children were suffering that awful mental fear that came when we had to be huddled in those boats. Major Butt was near me, and I know very nearly everything he did.

"When the order came to take the boats he became as one in supreme command. You would have thought he was at a White House reception, so cool and calm was he. When the time came he was a man to be feared. In one of the earlier boats fifty women, it seemed, were about to be lowered, when a man, suddenly panic-stricken, ran to the stern of it. Major Butt shot one arm out, caught him by the neck, and jerked him backward like a pillow. His head cracked against a rail and he was stunned.

"'Sorry,' said Major Butt, 'but women will be attended to first or I'll break every damned bone in your body.'

"The boats were lowered away one by one, and as I stood by my husband he said to me, 'Thank God for Archie Butt.' Perhaps Major Butt heard it, for he turned his face toward us for a second. Just at that time a young man was arguing to get into a lifeboat, and Butt had hold of the lad by the arm like a big brother and appeared to be telling him to keep his head.

"How inspiring he was. I stayed until almost the last and know what a man Archie Butt was. They put me in a collapsible boat. I was one of three women in the first cabin in the thing; the rest were steerage people. Major Butt helped those poor, frightened steerage people so wonderfully, tenderly and yet with such cool and manly firmness. He was a soldier to the last. He was one of God's greatest noblemen, and I think I can say he was an example of bravery even to the officers of the ship. He gave up his life to save others."

CHAPTER IX

THRILLING EXPERIENCES OF SURVIVORS

MARVELOUS BEHAVIOR OF MEN PASSENGERS — A SWEDISH OFFICER'S STORY — DISCIPLINE MAINTAINED TO THE END

❦⊰ FIRST WOMAN IN LIFEBOATS ⊱❦

rs. Dickinson Bishop, of Detroit, said:

"I was the first woman in the first boat. I was in the boat four hours before being picked up by the *Carpathia*. I was in bed at the time the crash came, got up and dressed and went back to bed, being assured there was no danger. There were very few passengers on the deck when I reached there. There was little or no panic, and the discipline of the *Titanic*'s crew was perfect. Thank God my husband was saved also."

P.D. Daly of England said he was above deck A and that he was the last man to scramble into the collapsible boat. He said that for six hours he was wet to his waist with the icy waters that filled the boat nearly to the gunwales.

❦ MEN PRAISED BY WOMAN ❦

One of the few women able to give an account of the disaster was Miss Cornelia Andrews of Hudson, N.Y. Miss Andrews said she was in the last boat to be picked up.

"The behavior of the men," she said, "was wonderful—the most marvelous I have ever beheld."

"Did you see any shooting?" she was asked.

"No," she replied, "but one officer did say he would shoot some of the steerage who were trying to crowd into the boats. Many jumped from the decks. I saw a boat sink."

Miss Andrews was probably referring to the collapsible boat which overturned. She said that the sinking of the ship was attended by a noise such as might be made by the boilers exploding. She was watching the ship, she said, and it looked as if it blew up.

❦ STORY BY SWEDISH OFFICER ❦

Lieutenant Hakan Bjornstern Steffanson of the Swedish army, who was journeying to this country on the *Titanic* to see about the exportation of pulp to Sweden, narrowly escaped being carried down in the sinking ship when he leaped out from a lower deck to a lifeboat that was being lowered past him. Henry Woolner of London also made the leap in safety. Lieutenant Steffanson thinks he made the last boat to leave the ship and was only about a hundred yards away when it went down with a sudden lurch.

He told about his experience as he lay in the bed at the Hotel Gotham, utterly worn out by the strain he had been under despite his six feet of muscle. It was also the first time he had discarded the dress suit he had worn since the shock of collision startled him from his chair in the café where he and Mr. Woolner were talking.

"It was not a severe shock. It did not throw anyone from his seat; rather it was a twisting motion that shook the boat terribly. Most of the women were in bed. We ran up to the smoking room, where most of the men were rushing about trying to find out what was the matter, but there was a singular absence of apprehension, probably because we believed so thoroughly in the massive hulk in which we were traveling.

❦ Sought to Calm Women ❦

"We helped to calm some of the women and advised them to dress and then set about getting them in boats. There seemed to be really no reason for it, but it was done because it was the safest thing to do.

"The men went about their task quietly. Why should they have done otherwise—the shock was so slight to cause such ruin. Mr. Woolner and I then went to a lower deck. It was deserted, but as we wished to find out what had happened we went down a deck lower. Then for the first time did we realize the seriousness of that twisting which had rent the ship nearly asunder. We saw the water pouring into the hull and where we finally stood water rose to our knees.

"Woolner and I decided to get out as quickly as we could and as we turned to rush upward we saw sliding down the portside of the drowning ship a collapsible lifeboat. Most of those it contained were from the steerage, but two of the women were from the first cabin. It was in charge of two sailors.

◄ Jumped into Swaying Boat ►

"'Let's not take any chances,' I shouted to Woolner, and as it came nearly opposite us, swinging in and out slowly, we jumped and fortunately landed in it. The boat teetered a bit and then swiftly shot down to the water. Woolner and I took oars and started to pull with all our might to get from the ship before she sank, for now there was little doubt of what would happen.

"We could see some gathered in the steerage, huddled together, as we pulled away, and then cries of fear came to us.

"We hardly reached a point a hundred yards away—and I believe the boat I was in was the last to get safely away—when the horrible screams came through the night and the ship plunged swiftly down. It was so terribly sudden, and then there was a vast quiet, during which we shivered over the oars and the women cried hysterically. Some of them tried to jump overboard and we had to struggle in the shaky boat to hold them until they quieted down.

◄ Victims Floated to Surface ►

"There was little widespread suction from the sinking ship, strange to say, and shortly after it went down people came to the surface, some of them struggling and fighting to remain afloat, and some were very still. But they all sank before we could reach them.

"It was bitterly cold and most of us were partly wet. It seemed hours before the *Carpathia* came up and took us aboard. Why, it was so cold that on board the *Titanic* we had been drinking hot drinks as if it were winter. The weather was absolutely clear, there was not the slightest fog or mist."

◀ BOILER BLAST SPLIT VESSEL ▶

Mrs. E. W. Carter left the *Carpathia* terribly shaken by her experience. She was met at the pier by Albert B. Ashforth. Mrs. Carter could not talk of the collision and the wreck, but Mr. Ashforth said that what had impressed her was the last boiler explosion.

"Mrs. Carter said that the shock of the collision was nothing," said Mr. Ashforth, "but the last boiler explosion tore the ship to pieces. She was in the last boat off."

What impressed E. Z. Taylor of Philadelphia most was the lack of excitement when the ship struck. He said he was on deck when the *Titanic* hit the iceberg and that he did not see any iceberg and did not think that anybody else did. Mr. Taylor said that he saw Mr. Ismay get into a boat fifteen minutes before the *Titanic* sank.

◀ BAND PLAYED UNTIL END ▶

Three sisters, Mrs. Robert C. Cornell, the wife of Magistrate Cornell; Mrs. E. B. Appleton, and Mrs. John Murray Brown of Action, Mass., went immediately to the home of Magistrate Cornell and related to George S. Keyes, a son-in-law of Mrs. Brown, what they had gone through. Mrs. Brown's story is the most vivid, as she left the *Titanic* in the last boat that got away safely.

"The discipline was magnificent. The band played, marching from deck to deck, and as the ship was engulfed you could hear the music plainly. The last I saw of the band the musicians were up to their knees in water.

"My two sisters and I were separated and each got in different boats. The captain stood on the bridge, and when the water covered the ship he was offered assistance and told to get in one of the lifeboats, but he refused to do so.

The eternal collision.

❮❮ Watched Parting of Astors ❯❯

"Mrs. Astor was in the lifeboat with my sister, Mrs. Cornell. I saw Colonel Astor help her into the boat. He said he would wait for the men. I saw him on the ship after our boat left the *Titanic*.

"We had a rough experience, many of the women having to use the oars. Mrs. Appleton's hands were badly torn, but I understand there was not a single case of illness among the survivors because of exposure.

Picture a situation such as this! Another woman and myself were waiting to be helped into the lifeboat. The woman held my arm. I do not know her name. There was just one seat left in the boat. The woman said to the men, 'This woman has children; let her go first. I'll take the next boat.' I believe she was put in the next boat. That boat was swamped."

❮❮ DROVE SEVERAL MEN BACK ❯❯

Mrs. Ada Clark, an Englishwoman who lost her husband in the wreck, stayed in her berth for half an hour after the collision.

"The shock was so light that it did not disturb me, and my husband told me to go back to sleep again. Then the stewardess came along and yelled, 'Everybody on deck.' There was no disturbance in filling the small boats. My husband put me in, kissed me goodbye and commended me to God. After I got into the boat two men tried to step in. An officer said that the boat was only for women and they stepped right back.

"I was in my night dress. The cold reached my brain and everybody in the boat was so benumbed from cold that we could not realize what a terrible thing had happened. Then somebody said, 'It's gone,' and we sat there without showing any emotion."

☀⊶ SAVED WITH HER CHILDREN ⊷☀

Mrs. Allen O. Becker, who is attached to the American Lutheran Missionary Society of Foreign Missions, and her three children, Ruth, eleven; Marion, eight, and Richard, six, were rescued from the *Titanic*.

She said she was awakened about 10:30 and a steward told her that everything was safe and that she could go back to sleep. In a half hour she was awakened by a steward who told her to take her three children in a hurry, as they were going to be put into a lifeboat. They did not get a chance to dress.

Mrs. Becker said that a steward took two of the children and she went with Ruth, but they all met in the same lifeboat. She said that they were in the boat until almost 5 o'clock when they were picked up.

☀⊶ JUMPED INTO SMALL LIFEBOAT ⊷☀

Abraham Hyman, a steerage passenger from Manchester, England, won his safety by leaving the steerage and going into the first cabin.

"I got alongside of a boat, and as they lowered it, full of passengers, I just crowded in beside the man at the tiller. They could have taken fifteen more people in our boat. There was no commotion in the first cabin. I heard that a man was shot in a panic in the steerage. When our boat got into the water it drifted under the exhaust of the *Titanic* and we were nearly swamped.

We rowed off for about half a mile and then saw the lights of the *Titanic* sink gradually out of sight. As the *Titanic* sank the lights went down, one after another."

Hyman said he heard of one man who had been sitting on a pile of deck chairs when the last explosion came who was blown off with the deck chairs. The man was found in the ocean on the deck chairs.

❮❮ BOILERS REND GREAT SHIP ❯❯

John Snyder and his wife of Indianapolis told how the boilers of the *Titanic* exploded and literally tore the ship to pieces.

"We were in our stateroom and I was asleep. The jar that came when the ship struck the berg did not even awaken me, and later when my wife aroused me we could hear persons running about the ship. Then a steward came and told us that there was danger and that we had better dress at once.

"We did dress and went on the second deck. There seemed no great excitement among the passengers, although the officers of the ship were giving orders to the crew to lower the lifeboats and were telling the passengers to get into them.

"We were told to get into a boat and we did, although at the time I much preferred staying on the *Titanic*. It looked safe on the *Titanic* and far from safe in the lifeboat. Before we knew what was being done with us we were swung from the *Titanic* into the sea and then the boat was so crowded that the women lay on the bottom to give the crew a chance to row.

❮⃛ *Titanic Sank Gradually* ⃛❯

"We went about 200 yards from the *Titanic*. We could see nothing wrong except that the big boat seemed to be settling at the bow. Still we could not make ourselves believe that the *Titanic* would sink. But the *Titanic* continued to settle, and we could see the passengers plunging about the decks and hear their cries.

"We moved farther away. Suddenly there came two sharp explosions as the water rushed into the boiler room and the boilers exploded.

"The explosions counteracted the effect of the suction made when the big boat went to the bottom and it is more than probable that this saved some of the lifeboats from being drawn to the bottom.

❮⃛ Explosions Killed Many ⃛❯

"Following the explosion we could see persons hanging to the side railings of the sinking boat. It is my opinion that many persons were killed by these explosions and were not drowned.

"Others of the passengers were tossed into the water. For an hour after the explosions we could see them swimming about in the water or floating on the life belts. We could hear their groans and their cries for help, but we did not go to them. To have done this would have swamped our own boat and everybody would have been lost. Several persons did float up to our boat and we took them on board.

"After we had got aboard the *Carpathia* we did not see J. Bruce Ismay until today, when he came on deck for a short time. He seemed badly broken up. You would hardly have known him."

❖ PERIL UNKNOWN AT FIRST ❖

A Mr. Chambers, one of the survivors, had this to say:

"The *Titanic* struck the iceberg. The passengers ran out, but, believing that the ship could not sink and being assured of this by the officers, again went back to their staterooms. After about two hours the alarm was sent out and the passengers started to enter the lifeboats. There was nothing like a panic at first, as all believed that there were plenty of lifeboats to go around."

After the lifeboat in which Mr. Chambers was had gone about 400 yards from the ship, those in it saw the *Titanic* begin to settle quickly and there was a rush for the remaining lifeboats. One was swamped. The great ship sank slowly by its head and no suction was felt by the boat in which Mr. Chambers was.

❖ GREEN LANTERNS SAVED MANY ❖

Henry Stengel of Newark said it was only the forethought of a member of a boat crew who was quick-witted enough to snatch up three green lights that saved a number of the lives of those adrift in the tiny lifeboat.

"These green lights," he said, "shining through the darkness enabled the other boats' crews to keep close together in the ice-filled waters."

Mr. Stengel put his wife in a boat and then followed. He said that early the next morning, shortly after they had been picked up, they saw floating far away a gigantic iceberg, with two peaks shining in the morning sun. There was the berg that sent the *Titanic* to the bottom, he thought.

⊰ JUMPED INTO SEA; PICKED UP ⊱

E.Z. Taylor of Philadelphia, one of the survivors, jumped into the sea just three minutes before the boat sank. He told a graphic story as he came from the *Carpathia*.

"I was eating when the boat struck the iceberg. There was an awful shock that made the boat tremble from stem to stern. I did not realize for some time what had happened. No one seemed to know the extent of the accident. We were told that an iceberg had been struck by the ship.

"I felt the boat rise and it seemed to me that it was riding over the ice. I ran out on deck and then I could see ice. It was a veritable sea of ice and the boat was rocking over it. I should say that parts of the iceberg were eighty feet high, but it had been broken into sections, probably by our ship.

"I jumped into the ocean and was picked up by one of the boats. I never expected to see land again. I waited on board the boat until the lights went out. It seemed to me that the discipline on board was wonderful."

⊰ SCENE AT RESCUE DESCRIBED ⊱

A passenger aboard the rescue ship *Carpathia*, Miss Sue Eva Rule, a sister of Judge Virgil Rule of St. Louis, Mo., detailed the thrilling scenes which marked the rescue of the survivors of the greatest maritime tragedy of the age.

"Unknown to the sleeping passengers, the ship turned abruptly to the north. None knew of the sudden change of course and the first intimation anybody got of the fact that anything unusual was about to take place was the order given the steward to prepare breakfast for three thousand.

"The tidings ran through the ship like wildfire and long before the Cunarder had come within the scene of the tragedy we were all on deck.

❦— First of Boats Sighted —❦

"Just as day broke a tiny craft was sighted rowing towards us and as it came closer we saw women huddled together, the stronger ones manning the oars. The first to come aboard was a nurse maid who had wrapped in a coat an eleven-months'-old baby, the only one of a family of five persons to be rescued.

"The men and women both seemed dazed. Most of them had almost perished with the cold, and some of them who had been literally thrown into the lifeboats perished from exposure.

"One of the most harrowing scenes I ever saw was the service of thanksgiving, followed by the prayers for the dead, which during the incoming of the little band of survivors, took place in the dining saloon of the *Carpathia*. The moans of the women and the cries of little children as their loss was brought home to them were heartrending. The hope that by some means their beloved ones would be saved never left the survivors.

❦ Survivors in Strange Dress ❦

"How those who were saved survived the exposure is a miracle. One woman came aboard devoid of underwear, a Turkish towel wrapped about her waist served as a corset, while a magnificent evening wrap was her only protection.

"Women in evening frocks and white satin slippers and children wrapped in steamer rugs were ordinary sights and very soon the passengers themselves were almost in as bad a plight as the rescued. Trunks were unpacked and clothing distributed right and left. Finally the steamer rugs were ripped apart and sewed into impromptu garments.

"My first view of the first boat sighted led me to think we were picking up the crew of a dirigible. Back of the boat loomed in the shadowy dawn the huge iceberg which had sent the *Titanic* to the bottom. The lifeboat looked like the usual boat which swings from a balloon.

❦ Women Discussed Scenes ❦

"After an hour or so of rest the only relief the women who had been literally torn from their husbands seemed to have was in discussing the last scenes. Shooting was heard by many in the lifeboats just before the ship took its final plunge and sank from sight, and the opinion of many was that the men rather than drown shot themselves.

"Mrs. Astor, who was one of the first to come aboard, was taken at once to the captain's room. Others were distributed among the cabins, the *Carpathia*'s passengers sleeping on the floors of the saloons, in the bathrooms, and on the tables throughout the ship in order to let the survivors of the wreck have as much comfort as the ship afforded.

"One woman came aboard with a six-months'-old baby she had never seen until the moment it was thrust into her arms as she swung into the lifeboat. Nothing could equal the generosity and helpfulness of *Carpathia*'s passengers."

❖⊷ DOUBTED WORD AT FIRST ⊶❖

Mrs. Louise Mansfield Ogden, of Manhattan, described tonight how she felt when she heard the *Carpathia*'s whistle sounding early in the morning. Mrs. Ogden asked her husband if there was a fog. Mr. Ogden had left the stateroom, however, and did not explain until some ten minutes later. The ship had then slowed down perceptibly, and Mrs. Ogden was pretty nervous.

Then her husband returned and told her that there had been a great accident and that the *Carpathia* was going to help.

"The passengers are asked to keep to their rooms," he said. "There isn't any need of being frightened. There's been no fire on our boat, but there has been an accident to the *Titanic*."

Mrs. Ogden thought that an accident to the *Titanic* was quite too ridiculous to think of and in that she shared the impression which, so she learned afterward, was current upon the *Titanic* after the latter had struck. Mrs. Ogden dressed hastily and went out on deck.

❖⊷ Boats Filled with Survivors ⊶❖

"I saw there on the bosom of the ocean a boat full of women and children. I suppose there must have been sailors there too, but I did not see them. There were only one or two women in evening dress, but most of them were clad in fur coats over their kimonos or nightgowns. They had on their evening slippers and silk stockings. Some of them wore hats.

"Far in the distance were two or three other black specks which we made out also to be boats. As daylight grew we made out more and more boats, three on one side of our ship and five on the other. Still later we picked up more.

"Here and there on the ocean's surface among the field of ice were bits of wreckage from the broken *Titanic*, and there were in sight many bergs eighty and ninety feet high. The passengers of the *Titanic* were taken aboard the *Carpathia* boatload by boatload up sea ladders.

❖ Most Women Hoisted Aboard ❖

"The women, most of them, were hoisted to the decks of the *Carpathia* in swings but a few were hardy enough to climb aboard by the sea ladders. The ocean all this time was calm as a lake and it was not a difficult task to take the excess passengers aboard.

"Some of the women helped out in the rowing in the lifeboats themselves."

Mrs. Ogden said that she saw the hands of Mrs. Astor, Mrs. John B. Thayer and Mrs. George D. Widener red from the oars. Most of the women were wet to the knees from the icy water that had slopped into the *Titanic*'s lifeboats.

CHAPTER X

SORROW AND HONOR
AND MEMORY EQUAL

HEROISM WAS UNIFORM AND UNIVERSAL AND
NO DISTINCTIONS NEED BE DRAWN

here are differences between the statements of those rescued from the perished *Titanic*. There are contradictions as well as differences. The fact, however, but confirms the sincerity and the endeavor to be truthful of all who try to tell the story. Agreement on every detail would suggest collusion and impair faith in what was said.

Readers who bear these facts in mind will get at the substantial truth of the various accounts and draw the correct conclusions from them. The one and great conclusion to be drawn is that which proves the bravery and unselfishness of officers and crew and passengers, the fortitude of women, the consideration of all for the children, and the credit the entire story casts on the unselfishness of human beings in a sudden and concerted exchange of worlds.

If the tragedy is sorrow's crown of sorrow, the tragedy is likewise a justification of the claim of the lost to the honor as well as to the pity of the race and to the assurances they were as dear to the Heart of God as they will forever be to the chronicles and traditions of men. Every soul alone knows and can never fully tell its own grief. Every household alone realizes and can never fully tell its own loss. No riven heart can ever believe another's heart suffers woe like unto its woe.

Equally true it is that there should be no comparisons instituted between exemplars of heroism where heroism was uniform and universal. Any one of us well knew friends who perished together, in one another's arms maybe. But others, too, know friends of theirs who met the same fate with the same courage. Comparison, contrast or competition of credit under such circumstances were revolting and impossible.

The men who have died for men have won the laurels of the race. The men who died for women are entitled to the love as well as to the laurels of the race. The men who died for little children are evermore shrined in the heart of Him "Who took the little children in His arms and blessed them," as He said, "For of such is the Kingdom of Heaven."

If there is any rose of distinction in the chaplet of memory, let it go to the husbands and wives who literally loved, lived and died together, each refusing to survive the other. For those dead the portals of Eternity swung wide open, but in the souls of those who went through them together must have been special joy, and for them well could be special honor and shall be.

The equal and equally honored and equally mourned dead should have and will have equal remembrance among the living. For them let sudden death be held to have been the assured glory of those who did die or were ready to die that others might live.

"For this cause shall a man lay down even his life," said He who once laid down even His for His enemies. In this instance not a few surrendered their lives even for strangers. The Friend and Father of all the race has no rebuke for those made in His image who followed His example. God accepts them. Christ receives them. Humanity cannot forget them. The summons all must answer, and most of us alone, is answered with special pathos and power on the sea, in the night and in grouped comradeship, with the consciousness and comfort as time recedes and Heaven opens, that is for them who live for others earth is well, for them who die for others Eternity has an abundant entrance into love ineffable.

❈⁃ THE LAST WORD FROM THE *TITANIC* ⁃❈

"We rowed frantically away from the *Titanic* and were tied to four other boats. I arose and saw the ship sinking. The band was playing 'Nearer, My God to Thee.'"

—Mrs. W. J. Douton, a survivor, whose husband was drowned

Nearer, My God, to Thee

Nearer, my God, to Thee,
Nearer to Thee!
E'en though it be a cross
That raiseth me;
Still all my song shall be
Nearer, my God, to Thee,
Nearer to Thee!

Though like the wanderer,
The sun gone down,
Darkness be over me,
My rest a stone;
Yet in my dreams I'd be
Nearer, my God, to Thee,
Nearer to Thee!

There let the way appear
Steps unto heaven;
All that thou sendest me
In mercy given;
Angels to beckon me,
Nearer, my God, to Thee,
Nearer to Thee!

Then with my waking thoughts,
Bright with thy praise,
Out of my stony griefs
Bethel I'll raise;
So by my woes to be
Nearer, my God, to Thee,
Nearer to Thee!

Or if on joyful wing
Cleaving the sky,
Sun, moon and stars forgot,
Upward I fly;
Still all my song shall be
Nearer, my God, to Thee,
Nearer to Thee.

CHAPTER XI

THE RESPONSIBILITY FOR FATAL SPEED

THE CAPTAIN WAS UNDOUBTEDLY CARRYING OUT INSTRUCTIONS OF THE OWNERS

he investigation of a committee of the United States Senate brought out all the material facts bearing upon the disaster that sent the *Titanic* and 1,595 persons to the bottom of the Atlantic. Mr. Bruce Ismay, managing director of the White Star Line, the first witness, deposed under oath that at the time of the collision the ship was not going at full speed. That is a matter of deduction from his testimony. "The ship's full speed was 78 revolutions. We did not make more than 72." The *Titanic* could steam between 22 and 23 knots an hour, so it is evident that her speed was at the rate of 21 knots, and therefore high in an ice drift where bergs could be seen by daylight and might be encountered suddenly after dark.

It was a clear, starlit night, the sea was calm, and except for the presence of loose floes and masses of ice with submerged bases there was no reason why the *Titanic* should not have been making good time. But the exception was very important. Obviously the great ship was proceeding at a high rate of speed under orders of the captain, who just as obviously was trying to carry out the instructions of his employers. If the *Titanic* was not as fast a ship as the *Lusitania* or the *Mauretania* she was expected to make a good record on her maiden trip, which could not be done unless she held to a prescribed route. It was certainly in the power of Mr. Ismay to have the *Titanic's* course changed to the south when dangerous ice was reported ahead. The warning had come by wireless from the *Amerika* the day before the disaster. But to take at once a more southerly course would have involved a loss in time of several hours at least on the maiden voyage of the great *Titanic*.

J.Bruce Ismay, White Star Manager.

After the tragic event it seems criminal that the course was not changed if the new ship was to be driven on at a speed of 21 knots. The alternative was to proceed slowly through the ice field, but at a rate to keep her under perfect control. A steamship of the size of the *Titanic* must maintain a speed proportionately greater than the speed at which a vessel of half her tonnage can be handled in an emergency. What, then, is the explanation of her forging through ice-strewn water almost at maximum velocity? Can there be any doubt that the risk was not understood? Swiftly to condemn is to lose sight of the fact that the experience of captains of transatlantic liners with fields of ice, particularly with bergs partly submerged, is negligible. To the commander of the *Titanic*, a veteran who had made the passage hundreds of times, the conditions that destroyed his ship presented no perils requiring him to slow down to headway speed or to safe maneuvering speed. It was sufficient for him that the night was clear, that the ice was loose. He believed, as he had declared before he took charge of the ship, that she was unsinkable. A faith fatal in its consequences, but he knew nothing of the power of a great mass of floating ice to tear out the side of a 45,000-ton ship and smash in her watertight compartments. It is clear enough that the loss of the *Titanic* and the sacrifice of two-thirds of her passengers and crew was due more to ignorance and misplaced confidence than to criminal carelessness.

After the event the world knows that a fearful risk was taken that ought to have been avoided. It is the old painful story of implicit faith in experience that proved valueless and in judgment that was fallible. A thousand and a half lives seem to have been wantonly sacrificed, but to place the responsibility without mitigation is not as simple as it seems in the shadow of the awful disaster. The verdict will be pronounced unflinchingly, but let the investigation be deliberate and evidence complete.

— *New York Sun*

The tragedy of the *Titanic*.

CHAPTER XII

OTHER CONTRIBUTING CAUSES OF THE DISASTER

IN ADDITION TO LACK OF LIFEBOATS, CREWS DID NOT KNOW HOW TO MANAGE THOSE THEY HAD — ALSO FIRE RAGED IN COAL BUNKERS FROM THE START — INEXPERIENCED CREW

 here was some criticism among the survivors of the *Titanic* crew's inability to handle the lifeboats. "The crew of the *Titanic* was a new one, of course," declared Mrs. George N. Stone of Cincinnati, "and had never been through a lifeboat drill, or any training in the rudiments of launching, manning and equipping the boats. Scores of lives were thus ruthlessly wasted, a sacrifice to inefficiency. Had there been any sea running, instead of the glassy calm that prevailed, not a single passenger would have safely reached the surface of the water. The men did not know how to lower the boats; the boats were not provisioned; many of the sailors could not handle an oar with reasonable skill."

☚ NO BOAT DRILLS HEAD ☛

Albert Major, steward of the *Titanic*, admitted that there had been no boat drills and that the lifeboats were poorly handled.

"One thing comes to my mind above all else as I live over again the sinking of the *Titanic*," he said. "We of the crew realized at the start of the trouble that we were unorganized, and, although every man did his best, we were hindered in getting the best results because we could not pull together.

"There had not been a single boat drill on the *Titanic*. The only time we were brought together was when we were mustered for roll call about 9 o'clock on the morning we sailed. From Wednesday noon until Sunday nearly five days passed, but there was no boat drill."

The White Star liner *Titanic* was on fire from the day she sailed from Southampton. Her officers and crew knew it, for they had fought the fire for days.

This story, told for the first time on the day of landing by the survivors of the crew who were sent back to England on board the Red Star liner *Lapland*, was only one of the many thrilling tales of the first—and last—voyage of the *Titanic*.

"The *Titanic* sailed from Southampton on Wednesday, April 10, at noon," said J. Dilley, fireman on the *Titanic*, who lives at 21 Milton Road, Newington, London, North, and who sailed with 150 other members of the *Titanic*'s crew on the *Lapland*.

"I was assigned to the *Titanic* from the *Oceanic*, where I had served as a fireman. From the day we sailed the *Titanic* was on fire, and my sole duty, together with eleven other men, had been to fight that fire. We had made no headway against it.

"Of course, sir, the passengers knew nothing of the fire. Do you think, sir, we'd have let them know about it? No, sir.

"The fire started in bunker No. 6. There were hundreds of tons of coal stored there. The coal on top of the bunker was wet, as all the coal should have been, but down at the bottom of the bunker the coal had been permitted to get dry.

"The dry coal at the bottom of the pile took fire, sir, and smoldered for days. The wet coal on top kept the flames from coming through, but down in the bottom of the bunker, sir, the flames was a-raging.

"Two men from each watch of stokers were told off, sir, to fight that fire. The stokers, you know, sir, work four hours at a time, so twelve of us was fighting flames from the day we put out of Southampton until we hit the iceberg.

"No, sir, we didn't get that fire out, and among the stokers there was talk, sir, that we'd have to empty the big coal bunkers after we'd put our passengers off in New York and then call on the fireboats there to help us put out the fire.

"But we didn't need such help. It was right under bunker No. 6 that the iceberg tore the biggest hole in the Titanic, and the flood of water that came through, sir, put out the fire that our tons and tons of water had not been able to get rid of.

"The stokers were beginning to get alarmed over it, but the officers told us to keep our mouths shut— they didn't want to alarm the passengers."

Another story told by members of the *Titanic*'s crew, was of a fire which is said to have started in one of the coal bunkers of the vessel shortly after she left her dock at Southampton, and which was not extinguished until Saturday afternoon. The story, as told by a fireman, was as follows:

"It had been necessary to take the coal out of sections 2 and 3 on the starboard side, forward, and when the water came rushing in after the collision with the ice the bulkheads would not hold because they did not have the supporting weight of the coal. Somebody reported to Chief Engineer Bell that the forward bulkhead had given way and the engineer replied: 'My God, we are lost.'

"The engineers stayed by the pumps and went down with the ship. The firemen and stokers were sent on deck five minutes before the *Titanic* sank, when it was seen that they would inevitably be lost if they stayed longer at their work of trying to keep the fires in the boilers and the pumps at work. The lights burned to the last because the dynamos were run by oil engines."

CHAPTER XIII

MORE OF THE TRAGEDY

DEATH WAITED FOR EVERY ONE, RICH AND POOR ALIKE, ON THE ILL-FATED SHIP

eorge D. Widener, the wealthy Philadelphian, and Arthur L. Ryerson of New York went to their deaths like men, is the statement by Mrs. Ryerson to her brother-in-law, E. S. Ryerson, after her rescue. She says that when the women were put into the lifeboats they saw Mr. Ryerson and Mr. Widener standing behind the rail of the *Titanic*, both waving their arms, throwing kisses and calling farewell to their wives and children. They believed there were boats enough for all. Mrs. Ryerson had her two daughters, Susan and Emily B., and a young son, John B., in the boat with her.

⋘ AIR-TIGHT CHAMBERS PROVED DEATH CELLS ⋙

That fifty or more steerage passengers of the *Titanic* were immured in a steel prison from which escape was impossible with the closing of the air-tight compartment doors in the steerage deck forward of midships was the statement made by a member of the ship's crew and who himself verified the fact that escape had been shut off for these unfortunates.

To have opened the doors which shut off these steerage passengers from the decks and possible escape would have been to shorten the life of the ship, he declared, and hurry disaster on all the hundreds crowded about the boat davits high above.

⟪ NO CHANCE FOR LIVES ⟫

"I know that fifty or more steerage passengers, whose quarters were on the same deck with the glory-hole used by the stewards of the second cabin, never got a chance for their lives. I know, because I nearly got caught myself by the closing of the water-tight doors leading from the working alley, which opened from the forward deck through to all the forepart of the ship.

"At the first shock all of the stewards in the my glory-hole, forty all told, tumbled from their bunks and went out through the working alley to see what the trouble was: I heard some one give an order, 'Look out for the water-tight doors.' A minute later I started to go back to the glory-hole to get a life belt, the order having been passed out to all members of the crew to equip themselves with these belts.

⟪ Steel Doors Slammed ⟫

"I could not get back through the alley to the glory-hole because the water-tight doors had slammed tight across the passageway. There was no way around it. There was no way for those on the other side of it, in the forespeak of the ship, to get out to open air.

"I know that none of the people from the steerage sleeping quarters beyond that water-tight door got out before it was shut, because they would have had to pass me in the alley, and none of them did. I spoke to one of the petty officers about the door being shut and all those people in there and he said: 'Well, what can we do about it now? If those forward compartments hold, then the air in them will keep us up all the longer.'"

◄◄ BELLBOYS AS WELL AS MILLIONAIRES ►►

Among the many hundred of heroic souls who went bravely and quietly to their end were fifty happy-go-lucky youngsters shipped as bellboys or messengers to serve the first cabin passengers. James Humphries, a quartermaster, who commanded lifeboat No. 11, told a little story that shows how these fifty lads met death.

Humphries said the boys were called to their regular posts in the main cabin entry and taken in charge by their captain, a steward. They were ordered to remain in the cabin and not get in the way. Throughout the first hour of confusion and terror these lads sat quietly on their benches in various parts of the first cabin.

Then, just toward the end, when the order was passed around that the ship was going down and every man was free to save himself if he kept away from the lifeboats in which the women were taken, the bellboys scattered to all parts of the ship.

Humphries said he saw numbers of them smoking cigarettes and joking with the passengers. They seemed to think that their violation of the rule against smoking while on duty was a sufficient breach of discipline.

Not one of them attempted to enter a lifeboat.

Not one of them was saved.

CHAPTER XIV

ODDITIES OF THE WRECK

FATE PLAYED SOME STRANGE FREAKS ALONG WITH THE HORROR — MONEY LESS VALUABLE THAN ORANGES

ne of the cabin passengers of the *Titanic*, Major A. G. Peuchen of Toronto, left more than $300,000 in money, jewelry and securities in a box in his cabin when he left the ship. He went back to his cabin for the box, but decided to take instead three oranges.

"The money seemed to be a mere mockery at that time," said the major. "The only trinket I saved was a little pin which I remembered had always brought me luck. I picked up the pin and three oranges instead of the money and documents."

Major Peuchen, who is president of the Standard Chemical Company of Canada and vice commodore of the Royal Canadian Yacht Club, was thrust into one of the boats by the captain and ordered to man an oar.

◄ DEMANDED A BATH ►

G. Wikeman, the *Titanic*'s barber, was treated for bruises. He declared that he was blown into the water by the second explosion on the *Titanic*, after her collision with the iceberg.

A passenger who was picked up in a drowning condition caused grim amusement on the *Carpathia* by demanding a bath as soon as the doctors were through with him.

◂◂ JUMPED FROM THE DECK ▸▸

Storekeeper Prentice, the last man off the *Titanic* to reach the *Carpathia*, swam about in the icy water for hours, but soon was restored. He said he had leaped from the *Titanic's* poop deck.

Mrs. James Baxter and her daughter, Mrs. P. C. Douglas of Montreal, Canada, when rescued were wearing the evening dresses that they had on at the Sunday night concert on the *Titanic*, having lost all their other wearing apparel.

◂◂ A ROMANCE OF THE WRECK ▸▸

In the midst of death and horror, Cupid played a little game and won. One of the girl survivors of the *Titanic*, Miss Marion Wright of Somerset, England, was married in New York the day after landing, to Arthur Woolcott of Cottage Grove, Ore. She came alone from her home in England to meet her fiancé and he had been in New York for nearly a week anxiously awaiting her arrival. The pair were schoolmates in England and became engaged before Mr. Woolcott left to become an Oregon fruit grower.

HYMN FOR SURVIVORS OF THE *TITANIC*
BY HALL CAINE

To the tune of "God, Our Help in Ages Past."

Lord of the everlasting hills,
God of the boundless sea,
Help us through all the shocks of fate
To keep our trust in Thee.

When nature's unrelenting arm
Sweep us like withes away,
Maker of man, be Thou our strength
And our eternal stay.

When blind, insensate, heartless force
Puts out our passing breath,
Make us to see Thy guiding light,
In darkness and in death.

Beneath the roll of soundless waves
Our best and bravest lie;
Give us to feel their spirits live
Immortal in the sky.

We are Thy children, frail and small,
Formed of the lowly sod,
Comfort our bruised and bleeding souls,
Father and Lord and God.

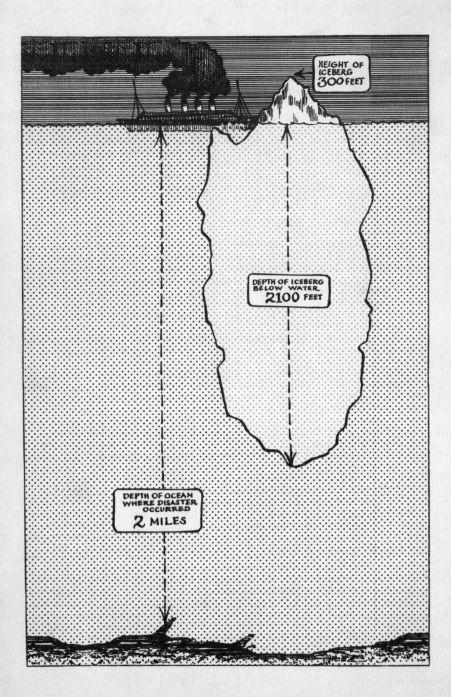

HEIGHT OF ICEBERG 300 FEET

DEPTH OF ICEBERG BELOW WATER 2100 FEET

DEPTH OF OCEAN WHERE DISASTER OCCURRED 2 MILES

CHAPTER XV

THE TERROR OF THE SEAS

BY FRED S. MILLER

here is one, and but one, danger to navigation against which the ingenuity of navigators is absolutely powerless, and this danger is formed by the vast icebergs—floating ice-prairies, some of them—which every month in the year, but more particularly in winter months, are sent in shoals from the Arctic and the Antarctic regions to float down the currents of the ocean until they are finally melted and mingled with warm waters. A brief account of the origin of these marine monsters, their action and the manifold dangers they present to sailors, will be of interest.

Greenland is the breeder of the iceberg for the northern seas. Greenland is a mysterious continent on which no vegetable life can endure. Its exact limits have never yet been traced, but is known to be comparatively flat, though covered to immense depths by snow and ice. This snow and ice forms constantly throughout the year, and has so formed since prehistoric times. It heaps up so that the surface of Greenland may be roughly compared to a vast hill. The enormous weight of this constantly forming ice causes movements of the masses from the center to the sea, and thus the glaciers are formed—vast processions of granite-hard ice which "flow" very slowly but irresistibly and for vast extents down to the water.

The size of these great moving plains is indeed almost unbelievable. The Humboldt glacier is sixty miles broad, its walls rise three hundred feet from the place where it meets the sea, and as to its depth inland it has been plumbed for half a mile. Every year it sends out over the ocean a mass whose area is greater than that of the State of New Jersey.

Another of the great Greenland glaciers, called the Jacobshaven glacier, is two thousand feet broad and one thousand feet high, and its output to the sea is estimated as being over 400,000,000,000 cubic feet of ice yearly.

Thousands of miles of this matter are constantly being emptied into the ocean, the rate of progress being about forty-two feet a day. Immense masses of solid ice creep along the shore, at the water's edge presenting a vertical face of steel-blue ice hard as flint, against which dash the angry waves of the Arctic. Out this ice pushes, day after day, until finally its own weight or the action of the water causes vast sections to break off with a roar like that of a thousand thunder claps and with a disturbance in the ocean that could only be compared to the commotion caused by the birth of a new island. Thus born, the berg floats gently down the currents for the Grand Banks of Labrador, where the fogs and mists that continually wreathe that region, shut the icy menace from view of the anxious mariner frequently until it is too late for him to turn his vessel to avoid them. In such weather it is of no help for the lookout on the tops that the iceberg frequently towers hundreds of feet into the air. It cannot be seen for the dense blanket of fog that shuts out sight and shuts out sound, so that even the wash of the waves dashing against the base of the approaching destroyer cannot be heard. Only by the cold radiated from it may its presence be guessed, but if the wind is blowing from the vessel to the berg the temperate cannot be felt lowering until the boat is so near that it is impossible to turn it before the crash comes.

Again, many of these great masses cannot be seen above the surface of the sea as they only extend comparatively a few feet into the air. Nevertheless eight-ninths of the berg is always under water, so that, especially at night, a vast plateau of ice may be gliding towards a steamer and giving no indication of its presence.

The steamer *Saale*, coming over the same course as that taken by the *Titanic*, was in 1890 subjected to almost the same experience although she escaped as by a miracle. Rushing along in the midnight gloom its path was suddenly barred by a black rampart of steely ice, 100 feet high. The lookout gave timely warning, the engines were reversed and the helm put hard aport, so that the steamer barely crunched along over the submerged foot of the berg, bumping heavily a few times and being shot off into deep water sidewise so that the coal and cargo were shifted. This listed the vessel heavily, in which plight she proceeded slowly to port, her starboard rail barely clearing the water.

The *Normania*, in 1900, had a similar experience. It turned just in time to avoid a direct impact with an immense berg, but it ran alongside of the floating mountain, shearing its sides and showering itself with ice scraped off, which loaded the decks.

But these are merely lucky escapes. By far the greater number of vessels, once they touch the frightful mass of beetling crag and jagged base, are lost on the moment of impact, the passengers being lucky if they have the time and the boats to escape with. The record of the sea is heavy with the account of gallant ships that perished, some with all on board. Many is the number that went down and were never heard from nor a vestige of them seen, but which were supposed to have been overborne by icebergs. Until very recent years the wireless telegraph was unheard of and ships suddenly overtaken could not communicate their plight but must vanish without leaving a record that they had ever been. In this way went the *Ismalia*, the *Columbo*, the *Homer*, *Zanzibar*, *Surbiton* and *Bernicia*, and to this day no light has been thrown on the mystery of their loss. Of course there are many more similar cases, any year of the last twenty being prolific with instances of these mysteriously disappearing ships.

The only vessel that can hope to escape destruction by contact with an iceberg is the especially strengthened ship built for Arctic exploration. Ships like the *Fram* of Amundson, or Peary's ship, are proof against even a head-on collision as they are very strong and very light. But an ocean liner is especially vulnerable. Going at the rapid speed that is nearly always maintained on these palatial ships, and with their enormous weight and displacement and their comparatively weak structure, the momentum which they acquire shatters them like glass when it is brought to an instant stop against a sluggish-moving mass say a mile long, two hundred feet above the water, 1,600 feet below the water, of a weight incalculably great and of a hardness like granite.

Much time has been spent and many efforts have been made to devise some instrument or discover some means whereby the presence of an approaching iceberg might be detected, but so far little progress has been made toward perfecting anything that at all answers the requirements. The towering berg can of course be seen for miles unless hidden by the fog, but what of the immense masses that lie scarcely visible in the water yet wholly destructive of whatever ship shall hurl itself upon that jagged floating reef of ice-coral? Ships that run on top of such bergs break literally in two, as their keels are not made to sustain a strain of balancing or "teetering" as the ship does when it runs upon the uneven surface of the berg.

But if icebergs are terrible they are beyond doubt among the most beautiful and superb manifestations of nature. Think of a mass of glittering minarets and towers, of domes, arches, colonnades, spires and special forms and features of its own uniquely beautiful—think of such a mass irradiant with a thousand variations of the rainbow hues and flashing in the sunlight of a northern summer day; think of a landscapeful of this delirious beauty, a bulk as large as the state of Rhode Island, moving majestically to the open ocean, breaking into mysterious peals of thunder as it dominates the sea! Perhaps it will receive and override some goodly vessel in its unruffled progress from the cold inconceivable which brought it forth. Perhaps the luckless voyagers will view its dreadful shape with an awe that will impel them rather to perish in the deep than to endeavor to seek refuge on the sheer and frigid walls that have o'erborne their ship. But presently the enormous edifice of ice itself shall sink and perish in the sea, merged with the enervating waters of the Gulf Stream— o'erborne as all things are and set to uses new by that emanation called by the learned "the opposition of forces" and by the wise called God, which keeps His ministering universe in equipoise and holds its balance true.

CHAPTER XVI

HEROES AT THE POST OF DUTY

DUTY AS STERN A MISTRESS AS DESPAIR GAVE MANY OPPORTUNITIES FOR THE DISPLAY OF BRAVERY — HELD PRAYER SERVICE AS SHIP SANK

he Rev. Thomas R. Byles, whose requiem mass was sung on Saturday at the hour he was to have officiated at his brother's marriage, was last seen leading a group in prayer on the second cabin deck of the *Titanic* when that ship sank. On the morning of the day the boat struck the iceberg Father Byles had preached to the passengers in the steerage and most of them knew him by sight.

When the *Titanic* struck the priest was on the upper deck walking back and forth reading his office, the daily prayers which form part of the duties of every Roman Catholic priest. After the real danger was apparent, survivors say Father Byles went among the passengers, hearing confessions of some and giving absolution. At the last he was the center of a group on the deck where the steerage passengers had been crowded and was leading in the recitation of the rosary.

This information was given by Miss Agnes McCoy, who was taken, as soon as she landed from the *Carpathia*, to St. Vincent's Hospital. She and her sister, Alice, were with their brother in the steerage. The two girls were put into a lifeboat and saw their brother swimming in the icy water. They called to him to get into their boat. He tried to grasp the side of the boat, but one of the sailors beat him back with an oar. In a minute one of the girls had reached the sailor and held his arms while the other sister pulled her brother aboard. Agnes McCoy gave this account:

> "I saw Father Byles when he spoke to us in the steerage, and there was another priest with him there. He was a German and spoke in that language. I did not see Father Byles again until we were told to come up and get into the boat. He was reading out of a leather bound book [his priest's book of hours] and did not pay any attention. He thought as the rest of us did that there wasn't really any danger. Then I saw him put the book in his pocket and hurry around to help women into the boats. We were among the first to get away and I didn't see him any more.

> "But there was a fellow on the *Carpathia* who told me about Father Byles. He was an English lad who was coming over to this country with his parents and several brothers and sisters. They were all lost. He was on the deck with the steerage passengers until the boat went down. He was holding to a piece of iron, he told me, and had his hands badly cut. One of the explosions threw him out of the water and he was picked up later.

> "He said that Father Byles and another priest stayed with the people after the last boat had gone and that a big crowd, a hundred maybe, knelt about him. They were Catholics, Protestants and Jewish people who were kneeling there, this fellow told me. Father Byles told them to prepare to meet God and he said the rosary. The others answered him. Father Byles and the other priest, he told me, were still standing there praying when the water came over the deck.

Nearer, my God, to thee.

"I did not see Father Byles in the water. But that is no wonder, for there were hundreds of bodies floating there after the ship went down. The night was so clear that we could see plainly and make out faces of those near us. The lights of the boat were bright almost to the last. They went out after the explosion. Then we could hear the people in the water crying for help and moaning for a long time after the boat went down."

⟨⟨⟨ STUCK TO THEIR POST ⟩⟩⟩

Postmaster General Hitchcock recommended that a provision be inserted in the pending Post Office Appropriation bill authorizing the payment of $2,000, the maximum amount prescribed by law for payment to the representatives of railway postal clerks killed while on duty, to the families of each of the three American sea post clerks who lost their lives on the *Titanic*.

"The bravery exhibited by these men," Mr. Hitchcock said, "in their efforts to safeguard under such trying conditions the valuable mail entrusted to them should be a source of pride to the entire postal service, and deserves some marked expression of appreciation from the government."

When last seen by those who survived the disaster these three clerks, John S. Marsh, William L. Gwynn and Oscar S. Woody, were on duty and engaged with the two British clerks, Iago Smith and E. D. Williamson, in transferring the 200 bags of registered mail containing 400,000 letters from the ship's post office to the upper deck. An officer of the *Titanic* stated that when he last saw these men they were working in two feet of water.

⫷ BAND KNEE DEEP IN WATER ⫸

Mrs. John Murray Brown, of Action, Mass., who with her sisters, Mrs. Robert C. Cornell and Mrs. E. D. Appleton, was saved, was in the last lifeboat to get safely away from the *Titanic*.

"The band played, marching from deck to deck, and as the ship went under I could still hear the music," Mrs. Brown said. "The musicians were up to their knees in the water when I last saw them. My sisters and I were in different boats. We offered assistance to Captain Smith of the *Titanic* when the water covered the ship, but he refused to get into the boat."

The names of the six Englishmen, a German and a Frenchman go down upon the roll of honor in the *Titanic* tragedy:

KRINS	CLARK
HUME	BRAILEY
TAYLOR	BREICOUX
WOODWARD	HARTLEY

In the list of second cabin passengers on the *Titanic*, the names of the eight are linked under the title of "bandsmen." When the last faint hope was gone, the eight musicians lined up on deck. Then solemnly and quietly the leader waved his baton, hands flew to instruments and over the ice laden water floated the strains of one of the most sadly beautiful hymns ever written. It was "Nearer, My God, to Thee." To their playing, more than 1,500 souls passed from life.

⫷ HEROISM AT HOME ⫸

Plucky Mary Downey at her switchboard played a hero's part as well as any of the rest of those of whom the *Titanic* disaster made heroes. During the days of suspense following the first news of the accident to the *Titanic*, while there were hundreds of people

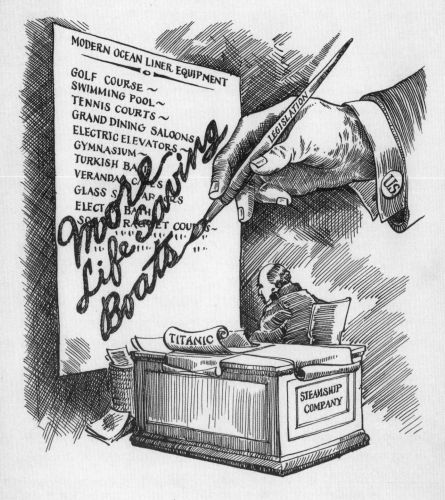

What was needed.

at the White Star offices and while even more were calling up either by local or long distance phone, one young woman sat at the White Star switchboard and bore the worst of everything. She was there to answer the first inquiries of relatives and friends of the *Titanic*'s passengers, to give them what hope she honestly could, to tell of the latest developments when there were any, to meet the quick demands of the officials of the line, and lastly to give immediate service to the throng of reporters that camped about the offices. All this, and more, too, when she had an unbelievably small amount of sleep.

The girl's name is Mary Downey. By the time the *Carpathia* had brought to port the remnants of the *Titanic*'s crew and passengers Miss Downey was as much of a hero among the White Star people as any one could be. A good many who were in a position to know have had much to say about her sticking to the job day and night. She was about the most composed person in the offices during those troubled times. Even Mr. Franklin, the general manager, took time once to remark to several reporters that Miss Downey "was a wonder."

The news of the disaster reached New York early on Monday morning; Miss Downey reached her place at 6 o'clock. She was there almost continuously until 8 o'clock on Tuesday morning. Part of the time she had an assistant. That afternoon after a few hours' rest, but without having gone to her home, she returned and again took up the answering of the endless inquires. The greater proportion of them she referred to clerks, but every one had to be looked into first by her. And when the clerks were all occupied she herself met the brunt of whatever words of sadness or criticism came over the wire.

Miss Downey had no idea how many phone calls she answered or made, but she knows that there are eleven trunk lines coming into the White Star offices, and that it was only during the hours of the night that these were not all in use.

The newspaper men who stayed around day and night appreciated as much, if not more, than any one else the service which Miss Downey rendered. For instance, when Mr. Franklin announced that the *Titanic* had sunk there were nearly a dozen reporters who rushed to office phones. Within a minute each had his own office on the wire and was flashing the news.

CHAPTER XVII

WILLIAM T. STEAD, SCHOLAR, DREAMER AND HUMANITARIAN

THE GREATEST AND MOST NOTABLE MAN ON BOARD THE TITANIC WHEN SHE SANK

orn in poverty, he rose by his natural genius for accomplishment to be a world power; always original, always independent, true to his ideals, first a worker and always a believer in good books.

"And you shall be kicked to death in the streets of London!" So said the clairvoyant. So prophesied the seer into the future, the gazer into the crystal bowl—but the crystal and the future were wrong.

W. T. Stead, world figure, human question mark, the man of the "automatic hand," who rose and rose and rose into the world from a beginning of nothing; a believer in the occult in the voices of the spheres, was not kicked to death in the streets. London did not see him die, as the clairvoyant had said. For it was in the whirling suction which followed the burial of the *Titanic*, in the rush and swirl and horror of fear dumbed death at sea that W. T. Stead went into that land from which in life he believed he drew messages.

And a figure of wonderful, almost grotesque interest went to death that night in the berg-ripped wreck of the steamer *Titanic*.

There was only one thing that could stop him in his course. That one thing was death. Prison could not stop his career, the anger of royalty could not check him, the deafening roar of a nation's displeasure caused him only to smile grimly and still keep on at what he had chosen to do.

He was born into the home of a Congregational minister on July 5, 1849, at Embleton, Northumberland, England. The father was a poor man. He had a large family. The boy's lot was a hard one. His child life was prosaic—and yet in everything, to him, there was a bit of a finer appeal, a wonderful yearning to find out the "why" of things, to know the reason for the being of this world, to remedy that which seemed wrong.

At fourteen entered poverty, stalking, ghastly poverty. A position was open as errand boy in a merchant's office. The salary was 4 shillings a week. All except 3 pence, or 6 cents a week, went toward the support of the family. The rest he could spend as he wished—6 cents!

Instead William T. Stead hoarded it, with a purpose, a goal that comes only from ambition. There were books to be bought; he must learn. He must study to satisfy the cravings of his mind. And so penny by penny the money was saved, to be disbursed now and then for some cheap edition of a book that was desired, a book that would be pored over and caressed and studied and absorbed in the hours of the night.

Then at seventeen came the great good fortune. The *Boy's Own* magazine was offering prizes for essays. Stead wrote one on Oliver Cromwell and a guinea came in payment. But the prize was not in money. It was in books, and perhaps that pleased the boy more than anything else could. Among the volumes that were selected by him was the poems of James Russell Lowell—and that volume was the making of William T. Stead's journalistic career.

In Russia, in Ireland, in Rome, in prison it was always

William T. Stead, sometimes called "the father of English journalists,"
who died in the sinking of the *Titanic*.

his prized possession. He carried it with him always, thumbed almost to pieces, underscored and marked in the margin.

In those "later years" the question of the unemployed came to Stead. It appealed to him. He pursued every account of work for the relief of the "out of works." He worked for the betterment of men who suffered through unemployment. He was made assistant editor of the *Pall Mall Gazette*, then virtual editor.

And it was then that William T. Stead began to wake up England.

London was rotten with a leprosy of white slavery. Nobles, members of Parliament, dukes, lords, all were in a great traffic in young girls that was carrying them down and down into a whirlpool from which there was no means of escape. London knew of the adder in its breast, but London tried not to see. London was aware that girls were being sold and bartered. But London did not seek further knowledge.

William T. Stead saw and knew. He looked farther into the future, and conceived what it would mean to bring all this to the surface, to expose it, and force it so visibly upon the people that they would demand action. The whole great reform might recoil upon the reformer and drown him in its tide of frenzied awakening.

The slothful morals might resist the efforts of the man who sought to arouse them and crush him. But Stead, grim faced, determined, decided to make the try. His name, his reputation, his freedom was on one side, against the torture of souls on the other. He might lose what he possessed in the effort to free the other, but then—

He took the dice box of fate into his hands and shook forth the cubes. They tumbled upon the green cloth of fortune, they wavered, then turned against him, then settled. He won.

And it was thus that they turned to the winning angle: Stead had found indisputable evidence of what he wished to prove.

He knew that power was against him, that money was against him, and that corruption was against him. He knew that the one element that could save him was the angry indignation of the populace. He set out to win that.

Bit by bit he gathered his testimony, name after name was secured, incident after incident was placed in sequence, and then, one morning in 1885, the blow that changed England's morals, or at least a part of them, fell. London awoke to stare, to gasp. Stead had called his exposure "The Maiden Tribute of Babylon," and there was truth to back every statement.

One exposure followed another, every fact was there, every bit of testimony stood forth in a nakedness of truth that was horrifying in its plainness. Stead was arrested. He was thrown into prison on the charge that he had committed an infraction of the laws. But he only smiled. He knew that he had won, that Parliament would be forced to pass a law that would wipe out the white slavery. And Parliament did.

Twenty years after visitors who went to the office of William T. Stead found him wearing a prison garb, numbered as it had been in the days when he suffered for the cause that he knew to be right. They were surprised. William T. Stead informed them that it was merely his anniversary, his means of celebrating a victory of the past.

It was in 1893 and 1894 that William T. Stead made himself best known to Chicagoans. For it was then that he came here, and, being impressed with Chicago and its inner workings, began the writing of a book that made him more famous than ever all over the world. The book was called *If Christ Came to Chicago*. When it was published it swept the country.

William T. Stead was, beyond doubt, the personal friend of more living and dead monarchs than any private citizen on earth. He interviewed more celebrities than any one writer in history.

CHAPTER XVIII

MANY MEMORIALS FOR TITANIC TRAGEDY

CHURCHES ALL OVER COUNTRY UNITE IN HOLDING SERVICES DEVOTED TO THE DISASTER — NEW YORK MASS MEETING

xpressions of tender, heartfelt sympathy for those who were in great grief; sorrow for those who died; glowing words of tribute for the heroism which had thrilled the world and then strong words urging legislation and regulation to prevent a recurrence of the *Titanic* catastrophe marked the memorial meeting at the Broadway Theater Sunday afternoon, April 21, 1912. Solemn as the occasion was, the great audience which jammed the auditorium from orchestra to top-most balcony could not forbear testifying its approval of that which was said at times or in joining in hearty approval of the resolutions which crystallized the sentiment.

The meeting was presided over by Frederick Townsend Martin and the principal speaker was William Jennings Bryan. Mr. Martin made a brief introductory address. The greater the sorrow the less the tongue could say, he declared; there are some sorrows too great to dwell upon. We can only mourn for those who perished; we can only sympathize with those that are suffering today. "We have rejoiced," he said, "over the great strides of business and commerce. We have believed in it, aided it until this commerce has grown too greedy and it has taken advantage of our confidence. It has preferred to spend its millions in extravagances and pennies for safety; we now reap the result of that policy."

Mr. Martin said that sorrow is a great educator. "We sometimes see further through a tear than through a telescope." It might be that out of this will come great good to the future. He called the conduct of those in the wreck heroic, showing a heroism "that only the angels can surpass, far greater than that shown on the greatest battlefield in the world's history." At the conclusion of his speech he introduced Mr. Bryan.

The epigram about seeing further through the tear than through the telescope had appealed to Mr. Bryan. "May we see through these tears now," he said:

"Our coming here today is an evidence that sometimes all of us can meet together, and we do meet together when drawn by a common purpose. There is a difference in education between us, much more than there should be, I fear; there is a difference in wealth, much more than there should be; there is a difference of church, much more than there should be, but we are all one when our hearts are touched, when we meet together upon the foundation of the heart."

Many more people had died in a given period than the *Titanic* catastrophe had called for, "it is not because so many died in a shorter period that we come here, but because of the suddenness of the death, the awfulness of it." Mr. Bryan used then the figure of a river and its tributaries. The storm of a single tributary had no effect on its volume; it is only when there is a general storm, when the water pours in from everywhere that the mighty stream rises, sweeps over its banks.

"So these people dying in a single moment have broken down all man made boundaries—we rush forth oversweeping everything that would prevent us.

"An occasion of this kind teaches its lessons. A great emergency is like a stage upon which the people play a part as before an audience. In the street you cannot tell the hero from the villain, but when you come upon the stage you see them all; they show us the little and the great, the rich and the poor, the wise and the simple as they really are; and this catastrophe has given us a chance to see how many heroes there are who only need a call forward to vindicate their right to be admired.

"I am proud of what we have learned of these men and these women, proud to know of their self-control that has given them the power to face death undismayed, aye, to stand back and say: 'Before me.'

"It is very easy to be polite when there is no danger in waiting; it is harder when delay, even for a moment, may mean death. I am proud of the records that have been made and glad that these illustrious examples come from every class.

"Some of the names are known. But it is not only they that need to be remembered at such a time as this. A gentleman was telling me yesterday a story he had heard from one of the survivors in that busy hour when all were seeking a means of escape. One of the passengers, a woman, was putting on a life preserver, and said to the steward: 'Where is yours?' The answer was: 'I am afraid there are not enough to go round.'

"He was doing what he could to save the others, and I am sure that none has read the story without being touched by it, of those wives who would not leave their husbands, who preferred to share the dangers of remaining with them to seizing the opportunity to escape. I knew one of these men in Congress. I was a colleague of Mr. Straus twenty years ago, and it is pleasant to know that he was a hero and not afraid; and it is sweet to know that the wife who had been his companion for so many years was true to the history of that earlier Ruth and preferred not to leave him— 'Entreat me not to leave thee.' These examples of manliness and womanliness are the heritage of our people. They make us proud of those whom we knew, who were a part of us.

"Nothing that we can say can bring back the dead. And little that we can say can soothe those who are under the shadow of a great personal loss."Those occasions are for the future more than for now, for others more than for ourselves."

Mr. Bryan told then of a conversation with a lawyer in a Western city years ago, who had said: "Without the shedding of blood there can be no remission of sins."

"He said," the speaker continued, "'You cannot correct a great wrong until somebody is killed; you may talk about dangers, but they will not listen.' Not until the tragedy of death shocks us will we pay attention. Often we do not know what needs to be done or provided until emergency throws its light upon the situation."

He told of his own experience in the West Indies last year when the ship upon which he was traveling ran upon a coral reef. The experience was not dangerous; there was no peril.

"But I learned then for the first time that they had but one wireless operator upon ships of that size and that by agreement the operators slept from 1:30 till 6 o'clock, four hours in the night when a sinking ship could not call another ship even if but a few miles away.

"The moment we found out the situation we were anxious that a law should be passed to require not less than two operators on a ship that there might be no delay in the securing of succor. We were not in danger and we could wait ten hours, but in three hours the *Titanic* went down. We learned we needed more operators and bills are now before Congress to remedy this, and I have no doubt that this great disaster, this greater, this gigantic, this titanic disaster will result in legislation that will be beneficial to those who come after.

"I venture the prediction that the wireless system will be made more immediately effective and efficient over a wider area and that the chance of danger will be diminished. I venture the assertion that as the result of the investigation now going on better preparations will be made with the lifeboats for the safety of passengers. I venture the assertion that less attention will be paid to comforts and luxuries that can be dispensed with and more thought given to the lives of those entrusted to the care of those shipbuilders and shipowners. I venture to assert also that the mania for speed will receive a check and that people will not be so anxious to get across the ocean in the shortest time as they will be to get across."

Mr. Bryan in conclusion referred to an old Greek game where the prize was to him that carried a lighted candle to a goal.

"And so these shipowners must learn that the race is not to the swift, but to those who can carry the light of life all the way over and not extinguish it on the way.

"I am glad to be one of this vast multitude to thus

testify by presence and word to the fact that we are all one in heart and feeling. I link my heart with yours in an expression of profound sorrow and in expression of deepest sympathy, and I link my hope to yours that this great, unspeakable disaster will bear a fruitage of good in larger safety to those who go down to the sea in ships."

Professor Felix Adler in his address to the Ethical Culture Society said in part:

"Heedlessness and culpable neglect brought on the *Titanic* disaster. The public in general must share the blame. It is pitiful to think of those golf links and swimming pools on the steamship which is now 2,000 fathoms deep. Though human weakness brought on the disaster sublime qualities were illustrated after its occurrence. The rule of the sea is based on moral equality of women and men. The statement made by some that women should have declined the preference well illustrates the rule. Inferior strength and less power of endurance is offset by a better chance for safety.

"There were places in the lifeboats for the physically weaker of the women and life belts for the physically stronger men. It has also been said that more valuable lives should have had preference, but those for whom this claim was made were the first to disdain it and they consorted with the undistinguished people in the steerage in the fine democracy of death.

"The most admirable feature was the calmness of those left behind."

The Rev. Dr. Charles A. Eaton said the following at the memorial service at the Madison Avenue Baptist Church:

"If the builders of the *Titanic* had had a real faith in the almightiness of God, they would not have believed that they could build something to master His seas. It was science they called upon, science, which since the days of Martin Luther has grown to be the mentor of the world. It gave them swimming pools, elevators, gorgeous suits and promenades, every comfort that a depraved and luxurious nation loves. When that proud ship sailed it had tortured the brains of the race in production and incarnated all of complex modern science. But science, which has brought the world between us and God, can never produce anything that will not crumble at the touch of God. That unconquerable boat went down.

"That one event has done more to dispel the wretched selfishness and sleepiness of our age than anything within my lifetime. With its best engines, its best staterooms, music, provender, diversions, its best people, it went down at a touch from God. We had forgotten the brooding deep and all that lies behind. We had not taken lifeboats.

"The managing director in his palatial saloon, the crew who did not drill, the man whose duty it was to bring up a bucketful of sea water for his thermometer and who filled it at the nearer faucet instead, all of them secure in their unsinkable ship—fools."

Twenty-two survivors from the *Titanic*, possibly more, attended the memorial service at the Cathedral of St. John the Divine Sunday morning after the disaster. Some of these survivors remained after the service to speak to the bishop and other clergy to thank them as they had in the service, they said, thanked God for their preservation.

At the bases of the chancel arch were great anchors of purple violets and upon the arches themselves were the British and American colors. Upon the fronts of the choir stalls were palm leaves and all doorways were draped in black and purple. The psalm from the burial service was sung while the people knelt, and the choir came in and at the close of the long and solemn service went out again in silence. The anthem was Sullivan's "Yea, though I walk through the shadow of the valley of death," and the prayers were from the same prayer book office for the dead.

The spirit of grief.

The Bishop of New York, the president of the house of deputies of the General Convention, the Archdeacon of New York, ex-President Smith of Trinity College and Canons Voorhees, Clover and Watson, with the cathedral dean, were among those who took part. So great was the number of people that they were seated in the choir stalls. Even then many stood. Bishop Greer's sermon was short, and near its close he bade the people pray and read a prayer for those in affliction, which brought the solemn occasion to its climax.

Maj. Archibald W. Butt, aid to the President, was educated at the University of the South, Sewanee, Tenn., and about twenty of his classmates, residents of New York and vicinity, attended a memorial service at St. Mark's Church, Second Avenue and Tenth Street. The Holy Communion service, a part of the memorial one, was especially for persons who knew the Major. Seventy-four came forward to receive it.

The rector of St. Mark's, the Rev. W. N. Guthrie, was a classmate of Major Butt and preached the sermon. His topic was "How Shall We View God in the Light of Such a Disaster?" After the service a committee was named to draw up resolutions of sympathy and forward them to the Major's family, which resides in Washington. Classmates who are members of the committee are Dr. John P. H. Hutchin, Beverly Wrenn, T. Channing Moore, Robert B. Elliott and William M. Puckette.

⁕ JEWS MOURN ⁕

Services in the Jewish temples of New York were occasions of mourning for the dead in the *Titanic* disaster. At several of the synagogues the catastrophe was the subject of the sermon.

At Temple Beth-El, Fifth Avenue and Seventy-sixth Street, of which Mr. and Mrs. Isidor Straus, who died loyally together, were members, all the representatives of the Straus family now in the city were present. They were ex-Ambassador Oscar Straus, brother of the dead philanthropist; Percy Straus, his son; Mrs. Percy Straus and her sisters, Mrs. Percy Straus' mother, Mrs. Abraham Abraham, widow of the late Mr. Straus' business partner; Mrs. Lazarus Kohns, his sister; Lee Kohns, his nephew, and Mrs. Edmund E. Wise, his niece.

The Rev. Dr. Samuel Schulman, who during his fourteen years incumbency as rabbi of the temple had been closely associated with Mr. Straus, could scarcely control his emotions as he spoke. He said in part:

"I knew Isidor Straus for fourteen years. He was a man with a great intellect, a sensitive conscience, a great heart, a loyal son of his people, and a loyal American—a great man.

"God's ways are not our ways. Therefore we should not attempt to define His motive in this tragic end of a great person. God sometimes, in His infinite wisdom, selects a man to designate that his life may be remembered by all mankind. At the conclusion of the Civil War it seemed to every one that the life of Abraham Lincoln was complete. His work, a great work, had been accomplished. Yet God saw one thing lacking. To perpetuate through the annals of time itself, one thing was essential. And God designated him and made a martyr of him.

"Isidor Straus was a great Jew. All the traditions of the Jew were dear to his heart. In the past we, as Jews, have been able to say the Jews are great philanthropists. Now when we are asked, 'Can a Jew die bravely?' there is an answer in the annals of time. When we are asked, 'What enabled Isidor Straus to do all these things?' our answers must be, 'God blessed him and gave him Ida Straus.' Isidor and Ida Straus were two persons with a single thought. Beloved and adored of each other in life, in death they were not separated."

At Temple Emanu-El, Fifth Avenue and Forty-third Street, the "Dead March in Saul" was played during the silent prayer. Sounds of sobbing filled the great edifice throughout the service, which was attended by Mrs. Benjamin Guggenheim, who was widowed by the *Titanic* catastrophe; Mr. and Mrs. Daniel Seligman, Mrs. De Witt Seligman, sister of Mr. Guggenheim; George Rosenheim, whose brother perished in the disaster; Mrs. Leo Greenfield and her son, Mrs. Edgar Meyer, the last three of whom were survivors of the wreck.

The Rev. Dr. Joseph Silverman preached the sermon:

"God is the Law Giver of the universe, and His laws are for the benefit of all, not of the few. When we violate the fundamental laws of nature we must suffer.

"Men learn by experience. Many may take comfort in the thought that the same errors will not again be committed, and that there will be no great sacrifice of life in the future from the same causes. All the progress in the world has been brought about by suffering on the part of individuals. Thousands have died and many more thousands have suffered in the cause of science. Millions have died on battlefields for the sake of liberty. Those on the *Titanic* when it went down must be added to the great roll of martyrs to progress."

❖ SPECIAL SERVICE FOR MAJOR BUTT ❖

President and Mrs. Taft attended services at St. Paul's Episcopal Church in Washington on Sunday in memory of Maj. Archibald W. Butt, the President's military aid, who lost his life in the *Titanic* disaster. Major Butt was a member of St. Paul's Church.

The services were held at 9 o'clock, before the regular morning service. Secretary of the Treasury MacVeagh, Secretary of War Stimson, Charles D. Hilles, secretary to the President, and many persons prominent in Washington society, including members of the Diplomatic Corps, were present.

The services were opened by the singing of "Nearer, My God, to Thee," the hymn which the heroic bandsmen on the *Titanic* played as the ship sank. The Rev. Frank Talbot, pastor of the church, took as the text of his sermon:

> "Greater love hath no man than this, that a man lay down his life for his friend.
>
> "It is not my purpose to dwell at length on the life, character and death of the gallant soldier who sacrificed his life for his brother men. This is not the place to speak nor to listen to human words, although we are here together in this little church, where our beloved friend was accustomed, as he said, to slip in from time to time to attend early communion service, with which his duties did not interfere, but we are here to listen to the words of that Man of Nazareth, who centuries ago died that men might live."

The Rev. Mr. Talbot indorsed the proposal to erect a monument to the memory of Major Butt.

"After all," he said, "length of days does not count much. It seems to me that had our friend lived to a ripe old age his influence for bravery and for nobility of character could not have been greater than it is today. His name and his valiant death will be treasured in song and story for centuries to come."

❧ WASHINGTON MOURNS ❧

The President also attended the regular services at All Souls

Unitarian Church, and in the afternoon went to the memorial services at St. John's Episcopal Church in honor of the Washington victims of the disaster. The Rev. U. G. B. Pierce, the pastor, referred to the *Titanic* disaster in his sermon.

"This is a memorial service," he said, "but during the last week our hearts have been so taxed, we have been strained with so many and so many conflicting emotions that the virtue of this service must be virtue of self-restraint. We have heard enough. We have felt too much and we are here now to drink anew at the fountain of life and to fan into flame the flickering torch of our faith. We need strength today in the face of this affliction."

The *Titanic* disaster was the topic of the sermons in many other Washington churches. The Rev. Samuel H. Greene, of Calvary Baptist Church, said:

> "In the events of the last week we have seen how sweet and beautiful womanhood could be and how noble manhood could be at its best, and we have seen how thin are the partitions that separate the fortunate from the unfortunate, the rich from the poor. It is not what a man has but what a man is that counts in the crisis of life.
>
> "On that night men stood back that women and children might reach a place of safety. The millionaire and the steward stood side by side and both alike were heroes.
>
> "But some one must bear the responsibility for that disaster through the years to come. So many went down, and they were not responsible for it. Let us wait patiently the result of a full and fair investigation."

In nearly all the Catholic churches of the city it was announced that requiem masses would be sung for the souls of the victims of the disaster.

❦ MANY MEMORIAL SERVICES IN CHICAGO ❦

Every seat of the auditorium was filled, and hundreds were turned away from the service in the Episcopal Cathedral because they were unable to gain an entrance. Rt. Rev. Theodore N. Morrison, bishop of Iowa, occupied the pulpit with Dean Sumner.

❦ Honor in Disaster ❦

"This is not a time for many words. Sentences are hollow and sentiments are commonplace and trite in the face of such an appalling disaster—disaster from the worldly standpoint, but an honor to God from the religious point of view.

"It has sobered the world. As we celebrate the death of little children as martyrs on Holy Innocents' day we will memorialize those who sank on the *Titanic* as the martyrs of this age sacrificed by God to arouse the world to a deeper spiritual realization, to a desire for a more splendid type and a consciousness that life is ever ending and we must be prepared to meet death when it comes.

"In the risen Christ we find promise of that life to come, not only for those who have gone before but for those who remain."

Special prayers were offered up for the dead and special music by the choir.

❦ Proof that Men Are Good ❦

Rev. Johnston Myers, of Immanuel Baptist Church, said:

"We may safely say that the *Titanic* was the most perfect human achievement up to the present time, the triumph of building on land and sea. In one hour last Sunday it was made a pitiful wreck, and in four hours the ocean closed over it forever.

"People are better than we think they are. Only a few months ago public opinion condemned as unfit one of the men who died as heroes and who is today acclaimed. The millionaires are not all bad men as it turns out.

"The nations are remembering God today as not before. People are praying this Sunday who did not pray last Sunday."

FALLOWS CONDEMNED OWNERS

Bishop Samuel Fallows, D. D., LL. D., rector of St. Paul's Reformed Episcopal Church, said:

"We cannot sufficiently condemn those in charge of the *Titanic* for dashing ahead in the face of danger of which they had been forewarned. But let us not forget that there is a well-nigh insane desire among us all for excessive speed, both on land and sea.

"It has been clearly demonstrated that in case of accidents provision is not made as to the number of lifeboats for caring for all on board any of the ocean lines. Must not this be remedied?"

BLOW TO CLASS PREJUDICE

Fredrick E. Hopkins, pastor of Park Manor Congregational Church, said:

"Among many lessons that we could learn from such a terrible calamity, one of the most important would seem to be this: That it ought to be for a long time more difficult than ever to arouse class prejudice, when this catastrophe has so clearly shown that the first and last thought the first cabin passenger had about the poorest woman in the steerage was that she should be given the first chance for her life no matter what happened to the man or woman of millions and of fame."

◄— GUILT AS OUR OWN —►

Rev. William E. Danforth, pastor of Christ Church, Elmhurst, said:

"In our dazed pondering of this *Titanic* disaster let us confess that the situation which shivered the ship shatters self-deluding ethics. The guilt is not that of any individual or corporation, but ours, in an age of mania for speed and smashing records. The one on whom to fasten the blame is every man to whom all else palls unless he rides in the biggest ship and the fastest possible. He will be guilty in his automobile tomorrow."

◄— DUE TO SPEED MANIA —►

Rev. W. H. Carwardine, pastor of the Windsor Park M. E. Church, said:

"Fifteen hundred human lives were sacrificed, sent to a watery grave, with the good ship *Titanic*, to satisfy the lust for speed, greed and the maritime supremacy on the sea of an Atlantic steamship company.

"The dare-devil insolence and pride of the human heart that would drive a vessel at such speed through a sea of ice and in spite of warning as to danger is staggering and incomprehensible."

◄◄┤ A THOUGHTLESS PEOPLE ├►►

Rev. Ingram E. Bill, pastor of the North Shore Baptist Church, said:

"The lust for conquest and a reckless disregard of human life is the glaring crime of the hour.

"What if the *Titanic* had evaded the icebergs and had swung into sight at the mouth of New York harbor hours before schedule time, smashing all the trans-Atlantic records?

"A thoughtless people, who now condemn the taking of a risk which resulted in the death of 1,500 precious souls, would have hailed with hysterical delight this new conqueror of the waves and yelled themselves hoarse in their demand for more speed and bigger and better achievements."

◄◄┤ REMEMBERED MANHOOD ├►►

M. M. Mangasarian spoke before the Independent Religious Society in the Studebaker Theater. He said in part:

"'Noblesse Oblige'—that glorious human precept was strictly observed by the splendid crew and passengers of the stricken *Titanic*. 'Be Britishers!' cried the veteran Captain Smith through a megaphone from his bridge. There is nothing more inspiring in any of the

Bibles in the world, except it be the more universal and thrilling challenge, 'Be men!' The *Titanic* episode has vindicated human nature grandly. Jew and Christian and agnostic forgot race and religion to remember that they were men."

CHAPTER XIX

STORIES OF THE RESCUED

SCORES OF FIRST-HAND ACCOUNTS REVEAL MORE OF THE
ACTUAL HAPPENINGS, THE BRAVERY DISPLAYED, THE ANGUISH
FELT AND THE DESPERATION OF THE SITUATION THAN DO
THE MOST GRAPHIC STORIES OF EXPERIENCED WRITERS

n the four days' cruise back to New York many
who had realized that their experiences would be
awaited by an anxious world put their stories on
paper while their nerves were still at tension from
the excitement of the disaster they had escaped.
Many others were interviewed on landing in New York or after
reaching their homes. While these accounts vary and conflict
often as to detail they point unanimously to the universal heroism
of crew and passengers that stamped the disaster with a new
character peculiarly its own.

◈⊰ MISS HIPPACH'S GRAPHIC STORY ⊱◈

"Yes, it was terrible. But it already seems like a dream to me."

So said Miss Gertrude Jean Hippach, daughter of Mr.
and Mrs. L. A. Hippach, of 7360 Sheridan Road, Chicago, when
questioned in regard to the frightful disaster to the *Titanic* and
its human freight.

Mrs. Hippach and her daughter left home the first week
in January to spend three months abroad, their object being
to improve Mrs. Hippach's health, and to visit relatives. Both

mother and daughter had been abroad several times, four times together. As Miss Hippach remarked: "It was my eighth voyage across the Atlantic; but I can't imagine I shall ever wish to cross it another time."

"We had expected to return by the Olympic, but found we were not allowing ourselves time enough for the short visit in Paris we had planned; so we engaged passage on the *Titanic*, which we boarded at Cherbourg. We touched at Queenstown and then turned toward America.

"The *Titanic* was so huge that it is hard to give an idea of it. It was over eight hundred feet, two blocks long, and wide in proportion. The staterooms were like rooms in a hotel. We had a regular bed and a handsome dressing table and chairs; and there was the lavatory with hot and cold water and there were electric lights and an electric fan, and an electric curling iron and of course push buttons—everything you could think of. One of our friends, when her husband asked her if she could think of anything to add to the equipment— laughed and said, 'Well, we might have butter spreaders; I can't think of anything else.'

"'Yet, there was no searchlight,' suggested a friend. My face was grave as I echoed in a low tone, 'No searchlight!'

"We had been to the concert in the evening till half past ten. The orchestra gave three fine programs every day; before luncheon, in the afternoon and after dinner every evening. They were all real musicians and were appreciated by the people on board, who were the finest lot of people I ever crossed with—people of leisure and good breeding, all of them.

"Well, we were asleep when the crash came; it was on our side and we awoke instantly and sat up in bed. Then the big boat shivered from the shock and then there was a long scraping, grating kind of noise and bumping, and then it was still.

"We ran out and found everybody out in the corridor, asking what was the matter. A steward came along and said it was nothing; we had only grazed an iceberg. He advised us to go back to bed. We went back. But Mother said, 'I've never seen an iceberg, and I'm going to put on some clothes and go on deck.' I tried to persuade her to go back to bed, but she was determined. I didn't want to be left alone, so I dressed, too. I was so sleepy it took me a long time to get dressed; but we both put on real warm clothes.

"If it had not been for Mr. Astor I believe we would have been among the lost. The last lifeboat was being lowered when Mr. Astor saw us. He ordered the boat raised so that my mother could get into it. 'Don't lower that boat until this woman gets in,' said Mr. Astor. We were compelled to climb through a porthole in order to reach the boat, but mother would not get into it unless I joined her. Mr. Astor again showed chivalry by pleading with the officers to permit me to get into the lifeboat, and they did.

"Colonel Astor was the calmest man during the exciting moments on the *Titanic* I ever saw. He smiled as he engineered the work of putting the women and children aboard the lifeboats. 'Don't worry, the *Titanic* will not sink, and we will all be saved,' said Mr. Astor, as he aided the frightened passengers into the boats.

"Well, we got into the lifeboat, though it didn't seem necessary, and it was so cold and so far down to the sea. But everybody was getting in. Ours was the last boat. Mrs. J. B. Thayer was in it. She rowed all night, hardly resting at all. She was so brave, although she must have known that her son and her husband— you know, she was the one who said her husband had 'better die than live dishonored.'

"And Mrs. Astor, too, was in our boat. We already knew her, that is, we knew who she was. She was crying and her face was bleeding from a cut. One of the oars struck her somehow. There was a little bride in our boat with her husband. She clung to him and cried that she would not go and leave him, so the officers finally pushed them both in together. There were about thirty-five in all in our boat, mainly from steerage."

In describing the lifeboat Miss Hippach indicated its length roughly as about thirty feet and explained that the air compartments were up just under the gunwale all around. She said that it was about five feet deep, with seats against the sides.

"We had gone back for our life belts before we got in, as the officers told us to do. I got mine on wrong side before and the officer changed it. That was the reason, perhaps, why some people couldn't sit down with them on. And we went back still another time and got some heavy steamer rugs, two of them, as the officers said it was going to be very cold on the water and we might have to stay out several hours. Even then we didn't expect the *Titanic* to go down, you see. The rugs were more than we needed, and we gave them to a poor woman who had on only a night gown and a waterproof coat and her baby was in a night gown only. That poor little baby! It slept through everything!

"After we had pushed away a little we looked at the steamer and I said to Mother, "It surely is sinking. See the water is up to those portholes!" And very soon it went under. To the last those poor musicians stood there, playing 'Nearer, My God, to Thee'. . ."

The girl's voice trembled and stopped.

"We had only one or two in the boat who knew anything about rowing and they kept turning it this way and that and again and again it seemed as if we might be capsized. But we did get away from the *Titanic* a little distance before it went down.

"We picked up eight men from the water, all third-class passengers, I think. The water was very still and the sky—so many stars! Nothing but the sea and the sky. You can't think how it felt out there alone by ourselves in the Atlantic. And there were so many shooting stars; I never saw so many in all my life. You know they say when you see a shooting star some one is dying. We thought of that, for there were so many dying, not far from us.

"It was so long, such a long, long night. At last there was a little faint light. The first thing we saw we thought was one of the *Titanic* funnels sticking out of the water. But it wasn't; it was the raft, the collapsible boat that didn't open, with twelve men on it, standing close together. They came up to us and demanded that we take them. But we thought they ought to say who they were; we were already pretty full and the water was getting rough. But they said they would jump in anyhow, so we let them come aboard, as we knew that jumping would surely capsize us. They were all stewards and waiters, men of the service of the *Titanic*. After we took them in it got still rougher, so that we sometimes shipped water. In fact, there was nearly a foot of water in the bottom of the boat and we hadn't a basin, or dripper, not so much as a cup to dip it out with. Meanwhile the waves were rising and if we hadn't been picked up when we were, another half hour would surely have been the end of us."

"How did you find things on the *Carpathia*?" she was asked. Miss Hippach exclaimed with enthusiasm:

"Just lovely. Nobody could have been kinder than they were. They kept their own people waiting and just took care of us. There was a warm blanket ready for each one and they had hot punch ready for us, or hot coffee and food.

"We couldn't sleep till night. We had to be crowded in somewhat. The passengers of the *Carpathia* gave up their rooms or shared them. We were with two old ladies who were very nice. But the first night we gave up our chance to two little brides who were very, very ill. They were from the *Titanic*. We slept on sofas in the dining saloon. The next night we had mattresses on the floor of the stateroom with the little brides and the old ladies slept somewhere else. The third night we slept in a regular bed."

Asked about the officers and servants of the ill-fated vessel, Miss Hippach said:

"They said cheerful things right through. You know they are under orders never to alarm the passengers, no matter what happens. So the stewardesses spoke soothingly, and assured us it was only a little accident, that we should all be coming back on board again in the morning, probably. But they knew, they knew they were lying."

❦ JACQUES FUTRELLE A HERO ❦

Mrs. May Futrelle, whose husband, Jacques Futrelle, the novelist, went down with the ship, was met by her daughter, Miss Virginia Futrelle, who was brought to New York from the convent of Notre Dame in Baltimore. Miss Futrelle had been told that her father had been picked up by another steamer.

Mrs. Charles Copeland of Boston, a sister of the writer, who also met Mrs. Futrelle, was under the same impression. Miss Futrelle and Mrs. Copeland, with a party of friends, awaited at a hotel the arrival of Mrs. Futrelle from the dock.

"I am so happy that Father is safe, too," declared Miss Futrelle, as her mother clasped her in her arms. It was some time before Mrs. Futrelle could compose herself.

"Where is Jack?" Mrs. Copeland asked.

Mrs. Futrelle, afraid to let her daughter know the truth, said: "Oh, he is on another ship."

Mrs. Copeland, however, guessed the truth and became hysterical. Miss Futrelle also broke down. She said:

"Jack died like a hero. He was in the smoking room when the crash came—the noise of the smash was terrific—and I was going to bed. I was hurled from my feet by the impact. I hardly found myself when Jack came rushing into the stateroom.

"'The boat is going down, get dressed at once!' he shouted. When we reached the deck, everything was in the wildest confusion. The screams of women and the shrill orders of the officers were drowned intermittently by the tremendous vibrations of the *Titanic*'s deep bass fog horn. The behavior of the men was magnificent.

"They stood back without murmuring and urged the women and children into the lifeboats. A few cowards tried to scramble into the boats, but they were quickly thrown back by the others. Let me say now that the only men who were saved were those who sneaked into the lifeboats or were picked up after the *Titanic* sunk.

"I did not want to leave Jack, but he assured me that there were boats enough for all and that he would be rescued later.

"'Hurry up, May; you're keeping the others waiting,' were his last words as he lifted me into a lifeboat and kissed me goodbye. I was in one of the last lifeboats to leave the ship. We had not put out many minutes when the *Titanic* disappeared. I almost thought, as I saw her sink beneath the water, that I could see Jack, standing where I had left him and waving at me."

❧ SAW ASTORS PART ❧

Mrs. Futrelle said she saw the parting of Col. John Jacob Astor and his young bride. Mrs. Astor was frantic. Her husband had to jump into the lifeboat four times and tell her that he would be rescued later. After the fourth time, Mrs. Futrelle said, he jumped back to the deck of the sinking ship and the lifeboat bearing his bride made off.

⋘ LADY DUFF-GORDON'S VIVID STORY ⋙

"I was asleep. The night was perfectly clear. I was awakened by a long grinding sort of shock. It was not a tremendous crash, but more as though some one had drawn a giant finger all along the side of the boat. I awakened my husband and told him that I thought we had struck something. There was no excitement that I could hear, but Sir Cosmo went up on deck. He returned and told me that we had hit some ice, apparently a big berg, but there seemed to be no danger.

"We were not assured of this, however, and Sir Cosmo went upstairs again. He came back to me and said:

"'You had better put your clothes on. Because I heard them give orders to strip the boats.'

"We each put on a life preserver, and over mine I threw some heavy furs. I took a few trinkets and we went up to the deck. There was no excitement at that time. The ship had listed slightly to starboard and was down a little at the head.

"As we stood there, one of the officers came running and said: 'The women and children are to go in the boats.'

"No one apparently thought there was any danger. We watched a number of women and children and some men going into the lifeboats. At last one of the officers came to me and said: 'Lady Gordon, you had better go in one of the boats.'

"I said to my husband: 'Well, we might as well take the boat, although I think it will be only a little pleasure excursion until morning.'

"The boat was the twelfth or thirteenth to be launched. It was the captain's special boat. There was still no excitement. Five stokers got in and two Americans, A. L. Solomon of New York and Dr. Stengel of Newark. Besides these there were two of the crew, Sir Cosmo, myself and a Miss Frank, an English girl.

"There were a number of other passengers, mostly men, standing nearby, and they joked with us because we were going out on the ocean.

"'The ship can't sink,' said one of them. 'You will get your death of cold out there in the ice.'

"We were slung off and the stokers began to row us away. For two hours we cruised around. It did not seem to be very cold. There was no excitement aboard the *Titanic*. We were probably a thousand feet away.

"Suddenly I clutched the sides of the lifeboat. I had seen the *Titanic* give a curious shiver. Almost immediately we heard several pistol shots and a great screaming arise from the decks. Then the boat's stern lifted in the air and there was a tremendous explosion. After this the *Titanic* dropped back again. The awful screaming continued. Two minutes after this there was another great explosion.

"The whole forward part of the great liner dropped down under the waves. The stern rose a hundred feet, almost perpendicularly. The boat stood up like an enormous black finger against the sky.

"Little fingers hung to the point of the finger and dropped into the water. The screaming was agonizing. I never heard such a continued chorus of utter despair and agony.

"The great prow of the *Titanic* slowly sank as though a great hand was pushing it gently down under the waves. As it went the screaming of the poor souls left on board seemed to grow louder. It took the *Titanic* perhaps two minutes to sink after the last explosion. It went down slowly without a ripple.

"Then began the real agonies of the night. Up to that time no one in our boat, and I imagine no one on any of the other boats, had really thought that the *Titanic* was going to sink. For a moment an awful silence seemed to hang over all, and then from the water all about where the *Titanic* had been arose a bedlam of shrieks and cries. There were women and men clinging to the bits of wreckage in the icy waters.

"It was at least an hour before the last shrieks died out. I remember next the very last cry was that of a man who had been calling loudly:

"'My God! My God!'

"He cried monotonously, in a dull, hopeless way. For an entire hour there had been an awful chorus of shrieks, gradually dying into a hopeless moan, until this last cry that I speak of. Then all was silent."

❦ CALIFORNIAN NOT ALARMED ❧

George Broden of Los Angeles, California, an athlete and head of a cement manufacturing concern, who was rescued by the *Carpathia*, said:

"I was preparing to retire when the crash came. It did not seem serious then. I put on an overcoat and went to an upper deck. Fifteen minutes later—there had been little excitement up to this time—a lifeboat was lowered. Shortly after this everyone rushed to the deck. Lifeboats were lowered on all sides.

"I was beside Henry B. Harris, the theatrical manager, when he bade his wife goodbye. Both started toward the side of the boat where a lifeboat was being lowered. Mr. Harris was told it was the rule for women to leave the boat first.

"'Yes, I know. I will stay,' Harris said. Shortly after the lifeboats left, a man jumped overboard. Other men followed. It was like sheep following a leader.

"Captain Smith was washed from the bridge into the ocean. He swam to where a baby was drowning and carried it in his arms while he swam to a lifeboat which was manned by officers of the *Titanic*. He surrendered the baby to them and swam back to the steamer.

"About the time Captain Smith got back there was an explosion. The entire ship trembled. I had secured a life preserver and jumped over.

"I struck a piece of ice and was not injured. I swam about sixty yards from the steamer, when there was a series of explosions. I looked back and saw the *Titanic* go down, bow first. Hundreds of persons were in the water at the time. When the great steamer went down they shrieked hysterically.

"When I jumped from the steamer into the water the band was still playing. The lights on the *Titanic* were lit until she sank.

"I was in the water two hours, clinging to a piece of wreckage when I was picked up by a lifeboat. About 6 o'clock in the morning the *Carpathia* appeared.

"I saw one of the stewards of the ship shoot a foreigner who tried to press past a number of women and enter a lifeboat."

❮❮ COUNTESS COMMANDS BOAT ❯❯

Miss Alice Farnam Leader, a New York physician, escaped from the *Titanic* on the same boat that carried the Countess Rothes.

"The countess is an expert oarswoman and thoroughly at home on the water. She practically took command of our boat when it was found that the seamen who had been placed at the oars could not row skillfully.

"Several of the women took their place with the countess at the oars, and rowed in turns, while the weak and unskilled stewards sat quietly in one end of the boat."

❮❮ LADY ROTHES' OWN STORY ❯❯

"It was pitiful, our rowing toward the lights of a ship that disappeared. We in boat No. 8 saw some tramp steamer's mast headlights and then saw the glow of red as it swung toward us for a few minutes, then darkness and despair.

"There were two stewards in boat No. 8 with us and thirty-one women. The name of one of the stewards was Crawford. We were lowered quietly to the water and when we had pushed off from the *Titanic*'s side I asked the seaman if he would care to have me take the tiller, as I knew something about boats. He said, "Certainly, lady." I climbed aft into the stern and asked my cousin to help me.

"The first impression I had as we left the ship was that, above all things, we mustn't lose our self possession;

we had no officer to take command of our boat and the little seaman had to assume all responsibility. He did it nobly, alternately cheering us with words of encouragement, then rowing doggedly. Then Signora de Satode Penasco began to scream for her husband. It was too horrible. I left the tiller to my cousin and slipped down beside her, to be of what comfort I could. Poor woman, her sobs tore our hearts and her moans were unspeakable in their sadness. Miss Cherry stayed at the tiller of our boat until *Carpathia* picked us up.

"The most terrible part of the whole thing was seeing the rows of portholes vanishing one by one. Several of us wanted to row back and see if there was not some chance of rescuing anyone that had possibly survived, but the majority in the boat argued that we had no right to risk their lives on the bare chance of finding anyone alive after the final plunge.

"Indeed I saw—we all saw a ship's lights not more than three miles away. For three hours we pulled steadily for the two masthead lights that showed brilliantly in the darkness. For a few minutes we saw the ship's port light, then it vanished, and the masthead lights got dimmer on the horizon until they too disappeared."

❦ APPEALS WERE IGNORED ❦

Mrs. Lucine P. Smith of Huntington, W. Va., daughter of Representative James Hughes of West Virginia, a bride of about eight weeks, whose husband was lost in the wreck, gave her experiences through the medium of her uncle, Dr. J. H. Vincent of Huntington, W. Va.

"The women were shoved into the lifeboats. The crew did not wait until the lifeboat was filled before they lowered it. As a matter of fact, there were but twenty-six people in the boat, most all women, when an officer gave instructions to lower it. Mr. Smith was standing alongside the boat when it was lowered. There was plenty of room for more people to get into the lifeboat, the capacity being fifty.

"Mrs. Smith implored Captain Smith to allow her husband in the boat, but her appeals were ignored.

"This lifeboat was permitted to be lowered with but one sailor in it, and he was drunk. His condition was such that he could not row the boat and therefore the women had to do the best they could in rowing about the icy waters.

"Mrs. Smith was in the third boat that was launched, and in that boat was Mrs. John Jacob Astor."

❖⤙ HER SON LEFT BEHIND ⤚❖

Mrs. Alexander T. Compton and her daughter, Alice, of New Orleans, were completely prostrated over the loss of Mrs. Compton's son, Alexander, who went down with the big liner.

"When we waved goodbye to my son, we did not realize the great danger, but thought we were only being sent out in the boats as a precautionary measure. When Captain Smith handed us life-preservers he said cheerily: 'They will keep you warm if you do not have to use them.' Then the crew began clearing the boats and putting the women into them. My daughter and I were lifted in the boat commanded by the fifth officer.

"There was a moan of agony and anguish from those in our boat when the *Titanic* sank, and we insisted that the officer head back for the place where the *Titanic* had disappeared. We found one man with a life preserver on him struggling in cold water, and for a minute I thought that he was my son."

CHAPTER XX

SURVIVORS' STORIES CONTINUED

VIVID RECOLLECTIONS AND PICTURES OF THE WRECK BY MORE OF THE RESCUED

rs. Turrell Cavendish, who was Miss Julia Siegel, daughter of the former Chicago millionaire merchant and clubman, Henry Siegel, was one of the survivors who landed from the *Carpathia*. Her husband was drowned. Mrs. Cavendish's father, Mr. Siegel, is part owner of Siegel, Cooper & Co. and is interested in a number of big ventures. Mrs. Cavendish is well known in Chicago society circles. Following is her story of the *Titanic*'s sinking:

"I was asleep when Mr. Cavendish awoke me and said the ship had met with an accident. I hurriedly put on a wrapper and one of my husband's overcoats and we both rushed to the upper deck.

"There were many persons there and the stewards were assuring them that the steamer was in no danger of sinking. They started to fill the lifeboats with women passengers when the cry, 'Save your lives!' rang out.

"I was in the second boat. My husband kissed me and bade me to remain in the boat, declaring he was all right. There was no light, but the sky was clear.

"Just as the lifeboat was lowered, I again kissed my husband.

"One man tried to get into the boat, but a sailor, after questioning him, threw him aside. A Canadian, who stated that he could row, turned to a group of men on the deck who were watching the proceedings and said:

"'I can row, but if there is room for one more let it be a woman. I am not a coward.'

"The women in the boat beseeched the man to row the boat for them, and those on the deck urged him to do so. With a parting handclasp he lowered himself by a rope to the boat and took his position there.

"I am prostrated by the loss of my husband, but rejoice in the fact that my two-year-old baby is saved, having been left at home."

◄ VIVID PICTURE OF WRECK ►

Miss Daisy Minahan, of Fond du Lac, Wis., who was with her brother, Dr. W. E. Minahan and his wife, told a graphic story of the shipwreck and the rescues. Dr. Minahan, she said, did his part in the saving of the women. Then with a farewell smile and the last words, "Be brave," to his wife, he went back on the deck, which even then was awash under his feet.

"There were more than a score of brides in our party on the *Titanic*, all coming back after their happy honeymoons abroad. We brought twenty of them, widowed by the terrible catastrophe, to homes of mourning and tears instead of happiness and bliss.

"We were sitting on the *Titanic*'s deck in the evening enjoying the crisp air and the starlit night. Old sailors told us the sea never had seemed so calm and glassy. About 9:30 o'clock the atmosphere took a sudden drop, which drove everybody inside the cabins.

"We must have been going at a terrific rate right in the direction of the icebergs, for the air became so chilly in a few minutes that we found it impossible to keep warm even when we put wraps and blankets around us.

"We had retired when there was a dull shaking of the *Titanic*, which was not so much like a shake as it was a slowing down of the massive craft. I noticed that our boat had come to a standstill and then we heard the orders of the captain and went on deck to see what it all meant.

"I never saw such composure and cool bravery in my life as the men of the first and second cabins displayed. Colonel Astor seemed to be the controlling figure. He, Major Butt, Mr. Guggenheim, Mr. Widener and Mr. Thayer clustered in a group as if they were holding a quick consultation as to what steps should be taken next.

"Then Colonel Astor came forward with the cry, 'Not a man until every woman and child is safe in the boats.'

"Many of the women did not seem to want to leave the vessel. Mrs. Astor clung to her husband, begging him to let her remain on the *Titanic* with him. When he insisted that she save herself, she threw her arms around him and begged him with tears to permit her to share his fate.

"Colonel Astor picked her up bodily and carried her to a boat, which was the one just ahead of ours, and placed her in it.

"I lingered with my brother and his wife, loath to leave them, although we all knew the ship was sinking and that the ocean would soon swallow up all that remained of the steamer. We both begged my brother to come with us, but he said: 'No, I will remain with the others, no matter what happens.'

"Then, when it was time to go, when the last boat was being lowered to the water line, we were hurried into it by my brother, who bade us goodbye and said calmly but with feeling: 'Be brave; no matter what happens, be brave.' Then he waved his hand and our boat shot out just in time to escape being borne down by the suction of the *Titanic*, as it went down.

"As the ship settled there was a terrific explosion, which rent it in two, and as it sank beneath the waves we could see my brother waving his hand to us, although it is hardly possible that he could see us, for none of us had a light. We had nothing except the clothes we had hastily donned. None of us had thought of putting provisions or water in the boats, for we knew the *Carpathia* had been signaled to come to our rescue and was on its way.

"We heard a number of shots as the boats were being lowered, but we were told it was the officers who were keeping the steerage passengers from stampeding into the small boats, which they repeatedly tried to do.

"There were no outcries anywhere except from the steerage.

"I shall never forget the calmness and quiet bravery that the men on board showed as they stood on deck and awaited the inevitable doom. Occasionally some of them would peer into the night toward our boats and wave at us. Then they would walk back to a group and everything would grow still again.

"I saw Guggenheim, Widener, Thayer and Ismay in conversation with Colonel Astor just after the ship struck the berg."

❊⤛ A MOUNTAIN OF GLASS ⤜❊

Thomas Whitley, a waiter on the *Titanic*, who was sent to a hospital with a fractured leg, was asleep five decks below the main saloon deck. He ran upstairs and saw the iceberg towering high above the forward deck of the *Titanic*.

"It looked like a giant mountain of glass. I saw that we were in for it. Almost immediately I heard that stokehold No. 11 was filling with water and that the ship was doomed. The watertight doors had been closed, but the officers fearing that there might be an explosion below decks called for volunteers to go below to draw the fires.

"Twenty men stepped forward almost immediately, and started down. To permit them to enter the hold it was necessary for the doors to be opened again, and after that one could almost feel the water rushing in. It was but a few minutes later when all hands were ordered on decks with life belts."

❊⤛ EXPERIENCE OF MRS. HENRY B. HARRIS ⤜❊

Mrs. Henry B. Harris, wife of the theatrical manager, who lost his life, told the following story:

"We were in our stateroom when the word was passed

for all passengers to put on life preservers and go on deck. This order followed within a few seconds after the ship struck. We did not realize the seriousness of the crash, thinking some slight trouble had happened to the engines. Even when the order was brought to us to put on life preservers and come on deck we still failed to realize the situation.

"As we went on deck we passed groups of men and women who were laughing and joking. When we reached the main deck, forward, and saw the lifeboats being swung overboard the seriousness of the matter began to dawn on us. Then came the command: 'Women and children first.'

"Officers and members of the crew went about repeating the words 'women and children first.' Many women had to be forced into the boats; some thinking it was a joke and others refusing to be parted from their husbands, fathers or brothers.

"When the passengers saw the seriousness with which the officers and crew of the *Titanic* went about their business they began to realize that something terrible had happened and began to make their way towards the lifeboats.

"Colonel Astor and Mrs. Astor were standing near us. When the men of the *Titanic* came to her and told her to get into a lifeboat she refused to leave her husband's side. Then I was asked to enter one of the boats. My husband told me to go but I did not want to leave him. He reassured me, saying the danger was not serious and that he would follow after me in a short time. Still I could not believe that everything was as he said. I felt that if I left him something terrible would happen. The officers told me I would have to get into a lifeboat. My husband told me to and finally I was led to the side and lowered into a boat.

"Mrs. Astor had left her husband and had been placed in another boat. As I was being lowered over the side I saw my husband and Colonel Astor standing together. Jacques Futrelle was standing near them. My husband waved his hand. That was the last I saw of him.

"For hours we sat freezing in the lifeboat. Then we saw the *Carpathia* and the men began to row in her direction. Then the *Carpathia* stopped and ropes were thrown to us and we were pulled against her side. Then rope ladders and swings were lowered and I was placed in a swing and pulled up to the deck. I stood watching the boats as they arrived and the passengers came on deck thinking every moment that my husband would appear. And then, when the last boat had been emptied I began to realize that he had gone down with the *Titanic*, which was nowhere in sight."

◀ HOW AN IRISH GIRL WAS SAVED ▶

When there was only one seat left in the last lifeboat of the *Titanic*, had Mrs. John Burke taken it the chances are that Miss Annie Kelly, a seventeen-year-old Chicago girl, might be at the bottom of the sea. So she told friends who gathered at her home to celebrate her lucky escape when the ship sank.

Miss Kelly told in a graphic manner the conditions in the steerage at the time the ship struck the iceberg and also how she was pushed into the last seat in the last boat.

With Miss Kelly when she arrived in Chicago was fifteen-year-old Annie McGowan, niece of Thomas McDermott, of Chicago, whose aunt, Miss Kate McGowan, perished in the lost ship. The girl was wrenched from her aunt's side and thrust into a boat, which pushed away from the ship. She never saw her relative again.

Annie Kelly and Annie McGowan embarked in the third cabin of the *Titanic* with the Burke family, which consisted of Mr. and Mrs. John Burke, who were coming to Chicago on their honeymoon; and Catherine and Margaret Burke, cousins of John and Margaret Manion, who were bound for Chicago to join their brother, Edward Manion. This is Miss Kelly's account:

> "I should not have been saved except for Mrs. Burke's refusal to leave her husband and the Misses Burke saying they would not go if their uncle and aunt could not go with them. I went in the very last boat and I was the very last passenger. The officer said there was room for just one more.

> "I was aroused by the call of the stewardess, who told us all to dress as quickly as we could, though she did not explain what was the trouble. I dressed and went up on the second deck. Annie McGowan was with me when I was going up the stairs, but she became separated from me at the head of the stairway, and was carried by the throng over to the other side of the ship. I did not see her again until I was on the *Carpathia*.

> "On the side where I was carried, some wild-looking men were trying to rush into boats, and the officers and crew fired at them. Some of the men fell. Others were beaten back by the officers, who used pistols on them."

⟪ TWO HEROIC CATHOLIC PRIESTS ⟫

Survivors of the *Titanic*, especially those from the steerage, told of the heroism of two Catholic priests who, after assisting women and children into the last boat, gathered about them the doomed passengers and calmly sought to comfort them in the face of approaching death.

The story of hope and faith evidenced in that hour by Father Byles of England and Father Peruschoetz, a German, entitles them to a high place in the roll of honor.

The two priests had held Sunday services in the morning and evening—for the Catholics of various nationalities, addressing them in German and English. The rosary and litanies had been recited by all.

The first news of the disaster brought the priests to the scene, where they joined with the other men in assisting to preserve order and insure the safety of the women and children. When men of all nationalities gathered about them and sought comfort and hope the two priests raised their voices and calmly, as if in the sanctuary, repeated over and over again the rosary.

No man, according to the story of those present, was turned away. The priests ministered to Catholics and non-Catholics alike. As the sinking vessel listed more and more the crowd about the priests grew larger, and all joined fervently in the prayers. Those in the boats pulling away from the vessel could see the men kneeling on the deck, but it is related that in the last moment, when the lights went out, no shrieks were heard nor cries of terror from the group where the faithful pastors serenely and devoutly sought to comfort those about them.

❦◄ VETERAN LAKE CAPTAIN WARNS ►❦

Another story of self-possession and undaunted courage in the face of death was that of Captain E. G. Crosby, of Milwaukee, veteran Lake Michigan navigator and president of the Crosby Transportation Company. "Better dress; all the other passengers are doing it," were his calm words to his wife and daughter as he entered their stateroom shortly after the collision. Captain Crosby was lost, but his wife and daughter were saved.

The majority of those who perished were caught sound asleep in their berths, according to Miss Crosby. The warning to his wife and daughter given, Captain Crosby hurried on deck to assist the other men. That was the last seen of him by Mrs. Crosby or her daughter. They were helped into the next to the last boat that left the vessel.

CHAPTER XXI

ON THE ROLL OF HONOR

SPENDID PUBLIC TRIBUTES TO WELL-KNOWN MEN AMONG THE HEROES OF THE TITANIC

⟪ ISIDOR AND IDA STRAUS ⟫

ho that hereafter writes of Isidor Straus can fail to write of Ida Straus? Linked in loyal life they were joined forever in a noble death.

If Isidor Straus was a great merchant, a great philanthropist, a clear-headed economist and a noble citizen, Ida Straus was a great woman, also a great philanthropist, a noble mother, a loyal, loving wife.

If Isidor Straus was the patriarch and honored head of a great family, Ida Straus was the serene and indispensable mistress of an honored home.

If Isidor Straus was a civic and commercial power, Ida Straus was a social and domestic force.

If Isidor Straus, after a life of honorable living, died a hero's death, so Ida Straus, after forty years of loyal loving, found of her own choice a heroine's end.

The beautiful examples of noble living and of nobler dying meet in these remembered names.

In an age of material absorption they have given a new and gentler illustration of the fidelity and tenderness of love.

In an age of domestic disloyalty and divorce they have wreathed a fadeless beauty around the deathless tie of marriage.

In life they were united. In death they refused to be divided.

As the world was better for their united living, so it shall be better for their loyal and undivided death.

⫷ MAJ. ARCHIBALD BUTT ⫸

In all the gallant band of men and gentlemen who went down to glory in the *Titanic's* wreck, there is none knightlier and more chivalrous than Archibald W. Butt.

He was a journalist, a gentleman, a courtier and a soldier in the armies of his country—measuring finely and fully to the high standards of each calling that he adorned.

It is not too much to say that even in the list of heroes in the epic of the sea there is a more than ordinary tenderness that wraps about the memory of the young chevalier of the new South—so gentle, so genial, so gifted, so tender and so true.

Born in Georgia of its bluest blood, Archibald Butt fought his way up like his fellows from the ashes of the South of the '60s—emerged from the ranks into dignity and high repute. As a Washington correspondent he was brilliant and popular. As a volunteer in the Spanish-American war he was a valiant and effective soldier in the ranks and as an officer. As the personal aide and social director of the White House he was the beloved of two Presidents of the United States, and won golden opinions from the American public.

And in the final supreme emergency—thinking always of others rather than himself, joining gentleness, serenity and firm authority with loftiest sacrifice—he mingled the finest pulses of his race and creed, and, wrapping the mantle of the English Sidney about his knightly shoulders, went down—to immortality.

Grieve not, the spirit of manhood still lives.

❈⊷ COL. JOHN JACOB ASTOR ⊶❈

The name John Jacob Astor, which has run for a hundred years through the commercial and social life of the metropolis, has taken on a new and nobler color in the passing of the last wearer of a famous name.

The last John Jacob Astor was a good soldier, a good sailor, an inventor of note, a builder of stately public houses, an author and a generous citizen. He was one among the few rich men of the metropolis who gave their money and themselves to the service of their country. He equipped a full battery of artillery and faced the bullets of the Spanish at Santiago.

One of the richest men in America, a leader of its ultimate social circle, newly married to a young and beautiful woman, John Jacob Astor had perhaps as much about him to make life sweet and to make death terrible as any man in all the great company of the *Titanic*.

And yet when the great moment came he laid down his life as bravely as a soldier, as calmly as a philosopher, and with as sweet and quiet a philanthropy as if his days were without color and his years without hope.

If the John Jacob Astors of the century past have lived like princes, this one but yesterday died like a man.

And the great name he bore is better known and better honored for his life and death.

The brave young wife who remembered others in mercy on that dreadful night has won the country's sympathy and respect.

❈⊷ GEORGE D. WIDENER ⊶❈

The Wideners of Philadelphia are a hearty race. Their money has not sapped their manhood. George D. Widener was big, red-blooded, genial—a man of courage and tenderness, so tried and proved that when the news came that there had been need for men to die on the *Titanic* in order that women and children might live his friends all knew that Widener was dead.

Men like J. Bruce Ismay may write voluminous statements and bring many witnesses to excuse their conduct, but the more they excuse, the more they accuse themselves. They can never answer the indictment of the men who die for the weak. Their clamor for exculpation is drowned by the deep silence of men like Widener.

It is recorded of George D. Widener that "he went down with the ship, fighting for the rights of women and children."

A hero is a man who actually does what everybody knows a man ought to do. To die for the women and children, in emergencies, when the issue is plain, is a matter of instinct with brave men. It is useless to argue against it—because it is not a theory.

It is a perception.

Widener understood. The man who could not resist an impulse to carry the clothes basket of an overburdened washerwoman understood perfectly.

❖ WILLIAM T. STEAD ❖

In his death as in his life, Mr. Stead stands dominant in the foreground of the greatest news. He was the *Examiner*'s staff correspondent in London—a master journalist, comprehending not only the outside of the news, but also its inner implications.

His eye was prophetic. He looked through events and beyond. He both made history and recorded it. If he had a generous dream of what ought to be, he was the first to help it come true.

Because of his passion for the improvement of the world, Stead was religious. He was on his way to America to preach a single sermon—and return. That sermon would have been preached Monday night, April 22, at Carnegie Hall.

Stead did not miss his engagement; the sermon was preached, indeed. It was flung to the world, with sublime persuasion—wireless, wordless—from the place where the *Titanic* went down.

For Stead was one of that group of immortals—of imperishable memory like the men of the Alamo—who would not leave the ship because there was no way to leave it with honor and humanity.

He died as he had lived—journalist, prophet, evangelist. Already his name was known everywhere; now his fame also is everywhere known—with a mourning affection that rejoices in the established greatness of his heart.

BENJAMIN GUGGENHEIM

It is related that the great Napoleon—as he sat on his horse observing a detachment of troops that were moving forward into the thick of a desperate action—called the attention of his aides to the pale, set face of a certain common soldier, saying:

"There is a brave man; for he knows his danger, yet faces it."

The stories of Benjamin Guggenheim's death do not say that he was pale or perturbed in the face of the great disaster; but they do say that he showed by his words and deeds that he knew his danger. Indeed, there is no other man in the long roll of *Titanic* heroes who left behind so clear a record of that consciousness of desperate peril which was Napoleon's test of perfect courage.

Whatever may be said of any other man, it is certain that Benjamin Guggenheim was not nerved to his deed of chivalry and sacrifice by any hope that the price would not need be paid. To

Johnson, his room-steward, whose superior prowess as a swimmer gave him an exceptional chance to be saved, Mr. Guggenheim said:

"I think there is grave doubt that the men will get off. Tell my wife that I played the game out straight and to the end. My duty now is to the unfortunate women and children on this ship. Tell her I will meet whatever fate is in store for me, knowing that she will approve."

— Chicago Examiner

CHAPTER XXII

COMMENTS OF THE PRESS

SOME OF THE PERTINENT EXPRESSIONS OF OPINION BY LEADING JOURNALISTS OF AMERICA

he trouble nowadays is that people wish to go with a rush. Subway trains whiz along through the tunnel at top speed; automobiles dash through the streets at a speed of a mile in two minutes, and ocean liners tear through the water, each striving to break a record. The *Titanic* was moving at a speed of twenty-one miles when she struck the iceberg which sent her down. So large and unwieldy was this ship that it could not be stopped inside of three miles. And yet it tore on through the night in the midst of ice fields. The passengers paid the penalty of speed. Not all the blame should rest on Captain Smith. It is not reasonable to suppose that he risked his own life, the safety of more than two thousand persons, and a valuable ship merely for the glory of making a record on a maiden trip. Not at all; Captain Smith went at high speed because every one was in a hurry; because the persons on the vessel wished to get to New York as soon as possible. The speed was deadly; and there is a lesson in this awful shipwreck. Do not rush when rushing imperils life.

— *Morning Telegraph,* New York

❦ THE ILIAD TURNED EPIC NOW ❧

The seas have been swept by an epic that will live while the memory of man endures.

The world has had a new baptism of heroism and splendid sacrifice, and the race of men is consecrated anew by sublime example to chivalry and unselfish faith.

It comes timely to a carping age, this message of denial which the remorseless sea sends above its engulfing billows to this old world, said to be sordid, and thought to be hard and cold.

There were no distinctions of race or creed or culture in the altruistic heroism which from the sinking decks of the *Titanic* enriched history and inspired the world.

There stood the splendid Englishman at the wheel and there stood the splendid Americans on the deck. The stanch Catholic, the loyal Protestant, the gentle Hebrew, and even the gambler, without creed, mounted the heights of godlike heroism before they went to death in the sea.

Captain Smith was born in Surrey, Colonel Astor was born in New York. Isidor Straus, a great Jew, wrapped his arms about Ida Straus, a great Jewess, and they went smiling down to death together. Colonel John Jacob Astor was a man of millions, which are said to make men cold. He was a type of fashion, a master of cotillions, and a leader of the 400 in the brightest city in the world.

William T. Stead was a man of letters, a pale, patient student, in whose thoughtful veins the red blood of resolution might have been expected to go slowly. Henry B. Harris was a playwright and a master in the mimic world, where life's passions and splendors are said to be unreal. And Archie Butt was born of the chivalric South, cavalier in manner and gallant in speech— the velvet-gloved and iron-handed Archie—perhaps the gentlest and knightliest soul of all that hero band.

"For there was neither East nor West
Border nor breed nor birth
When these brave men stood face to face,
Though they came from the ends of the earth."

So that it was the race—the race of men who have blazoned in light and glory against the aurora of that solemn dawn, the inspiring, the glorious fact that neither greed nor gold, neither ambition nor power, neither fashion nor folly have corrupted or crushed the indestructible chivalry and sacrifice that lives in the hearts of men.

Take heart, oh doubter, and let cynic and skeptic go henceforth slow. The race is not degenerate, and the future of our country is secure. The *Titanic*, sinking, uncovered the universal heartbeat that can always be reached by life's noblest appeal.

To protect the weak and to love your neighbor as yourself is the highest Divine and human law condensed through a thousand years of living.

In this high conception the stupendous incident may reach its noblest meaning. The *Titanic's* heroes have not died in vain. It was worth the majestic steamship, and even worth two thousand human lives, if the world comes once more to believe in its better self—if the race is inspired and led to better living and to better dying—to greater charity and to nobler hope.

And so this vast iliad of the ocean may soften at last into the most serene and splendid epic ever writ on land or sea.

— John Temple Graves, *Chicago Examiner*

Newspaper boy Ned Parfett sells copies of the *Evening News* telling of the *Titanic* maritime disaster, outside the White Star Line offices at Oceanic House in London's Cockspur Street, April 16, 1912.

❧ REGULATION OF WIRELESS REQUIRED ☙

America may make, as the London papers have said, "hasty and often cruel verdicts," but in the *Titanic* case America is becoming daily more glad that the investigating committee of United States senators had the energy and vision to board the *Carpathia* before she docked. Else, who knows how little of the truth about the wreck we would ever have known?

The testimony has taught us that even the wireless, the wonderful instrument for lessening the perils of the sea, may become in unworthy hands an instrument for capitalizing human agony instead of alleviating it. We have learned that this new force must be sternly regulated if it is to perform its due service to humanity.

— *Chicago Evening Post*

❦ MEN, WOMEN AND CHILDREN—THESE THREE ❦

We call our age commercial, material. In a sense it is. But we are apt to carry our meaning far. Especially as regards women we imply that chivalry is passed. "A gentleman of the old school," we say. Our epithets of courtesy are taken from the Middle Ages.

Of late years, with women among the workers, the keen edge of gallantry, we say, is lost. With suffragists demanding equal rights, there has been lament for the good old days of "woman's sphere" and man's gentleness in power.

And now—

"Women and children first!"—on the listing deck of the *Titanic*.

Stoker, valet, millionaire, responded true to the primal instinct—true, too, to the finest culture. Stories there are (probably true) of some frenzy, of some unmanliness. Let them pass. Cowards were of the brave Stone Age. Cravens were a reproach to knighthood. The large fact stands undimmed—women and children were the first care. Not many women were lost save by some act of devotion on their part, or some mischance. Few men were saved except by some good chance, or some rare fortitude.

The greatest sea tragedy of history is in the material twentieth century. More sacrificial idealism relieved it than any recorded incident of the Golden World affords.

We may cherish that and build high hopes on it. We may cherish it for what it means for the women and children of the race. Man still has the patriarchal impulse to protect his womankind. A tremendous incident disclosed it in tragic beauty. Less dramatically, the same impulse has shown itself as clearly to hearts of faith.

A civilization whose men of all individual types stand back from the lifeboats for the women and children is only superficially material. What of neglect and cruelty oppress its women and children will not endure.

It is written:

"Greater love hath no man than this, that a man lay down his life for his friends."

The sacrificial love of the man race for the woman race, the child race—that endures.

— *Kansas City Star*

❄ FROM THESE HONORED DEAD ❄

Most of the dead on the *Titanic* died heroically, yielding their lives both that the women and children of the ship's company might live and that the lives of thousands of others totally unknown to them might be spared in the future. They perished for their fellows as truly as soldiers who give their lives in a nation's defense, for the world can never forget what they did and suffered in the supreme crisis, and will be made wiser and better for their inspiring sacrifice.

It is a painful thought that some must die that others may be saved and many suffer that a succeeding generation may benefit. But that is the law of this imperfect world, slowly struggling toward distant goals of a moral and material betterment. Progress can seldom be accomplished without the martyrs whose sufferings stir the public imagination and set at work the influences which compel another forward movement. It is for the living always, as Lincoln said at Gettysburg, to take increased devotion to the cause for which the dead have given the last full measure of devotion. The heroes of the *Titanic* will not have died in vain if by their sacrifice the perils of the sea are henceforth materially lessened and the recklessness with which those perils have been faced becomes a discreditable memory.

—*New York Tribune*

❖ WOMEN AND CHILDREN SAVED ❖

After the world had settled down to the belief that no lives had been lost by the accident to the great ocean liner, the *Titanic*, it learned with horror that more than 1,500 of the passengers and crew went to the bottom of the Atlantic in that ill fated vessel.

The shock of this terrible loss is accompanied by feelings of pride and admiration because the men on board, facing death, stood back and gave the women and children the places in the boats that were launched as the big ship settled down into its grave. There were heroes in plenty on board the *Titanic*, as well as men of great wealth and wide renown.

The human race mourns its heavy loss, but it accepts the boatloads of rescued women and children as a precious token of the high courage and the loving self-sacrifice of the men who took the plunge to the bottom of the deep that the weaker companions of the peril might live. Greenland's glaciers, which in Melville Bay and elsewhere expose at the water's edge sheer fronts of ice having a width of twenty-five to thirty miles, calved the icebergs that thronged the pathway of ocean vessels in the North Atlantic.

While the old deadly perils still haunt the sea lanes— perils that, as the unhappy *Titanic* has demonstrated, even the greatest ships cannot face with safety—the wireless is now available to summon help in any time of calamity. It is a bitter disappointment to learn that aid promptly extended did not suffice to save many hundreds of those on the *Titanic*. The one bit of consolation from the calamity is that the world has been enriched by another example of tender devotion to others on the part of men who were facing imminent death.

— Chicago Daily News

❊⊸ NO HERO DIES IN VAIN ⊸❊

For the rest of the world, for the millions whom the disaster did not touch personally, the lasting thought will be this:

Every great disaster, every great affliction, rightly interpreted and rightly used, is a lesson and a help to all of the human race throughout the future.

No martyr, no hero, dies in vain. The safety and the progress of the world are built upon the afflictions and the sufferings of those that have gone before us.

The children of the men and women that died on the *Titanic* will find the last expression of their duty in Lincoln's immortal words of dedication upon the battlefield of Gettysburg:

"We have come to dedicate a portion of that field as a final resting place for those who here gave their lives that the nation might live. It is altogether fitting and proper that we should do this. But, in a larger sense, we cannot dedicate—we cannot consecrate—we cannot hallow—this ground. The brave men, living and dead, who struggled here have consecrated it far above our poor power to add or detract. The world will little note, nor long remember, what we say here, but it can never forget what they did here. It is for us, the living, rather, to be dedicated here to the unfinished work which they who fought here have thus far so nobly advanced. It is rather for us to be here dedicated to the great task remaining before us—that from these honored dead we take increased devotion to that cause for which they gave the last full measure of devotion—that we highly resolve that these dead shall not have died in vain."

Life is one great battlefield. This earth has been a field of battle through all the thousands of centuries of life here. And for many centuries to come it still must remain a field of battle.

Those that survive must find their comfort in the heroism of the dead. And the race must find its lesson and its growth in the experiences and the suffering of the past.

Far out in the Atlantic Ocean there is a dreary spot, with here and there, perhaps, a broken oar, or a floating body. Desolate and wide the ocean spreads beneath the dark sky, at the spot where the great ship sank.

But in all space that ocean and the planet upon which it rolls are but a speck.

Time is the real ocean, the ocean that has no limit to its depths and that has no boundaries.

The brave men and women of the *Titanic* are added to the heroes of that great ocean of time—the ocean that covers all the past, the ocean beneath whose waves brave men and women lie at rest, all the brave spirits that have lived honorably and died courageously on this planet.

It is a glorious thing for a man or a woman to have his name added to the list of those consecrated by time and by courage.

Every noble death does its good work. Other human beings will travel more safely and many thousands of lives will be saved as a result of the disaster so needless, so cruel.

— *Chicago Sunday Examiner*

❧— FROM CAPTAIN SMITH'S WIDOW —❧

The widow of Captain Smith, commander of the *Titanic*, wrote a pathetic message which was posted outside the White Star offices in London on the Thursday following the wreck. It read as follows:

"To My Poor Fellow Sufferers:

My heart overflows with grief for you all and is laden with sorrow that you are weighed down with this terrible burden that has been thrust upon us. May God be with us and comfort us all.

Yours in deep sympathy,
Eleanor Smith"

❧— FACTS ABOUT THE TITANIC —❧

The *Titanic*'s length over all was 882 feet 6 inches. 182½ feet more than the height of the Metropolitan Tower in New York City, and 3 1/3 times the height of Chicago's highest building. The Bunker Hill monument is one-fourth as high, and the Washington Monument itself 300 feet shorter.

Some of the statistics follow:

Tonnage, registered	45,000
Tonnage, displacement	66,000
Length over all	882 feet, 6 inches
Breadth over all	92 feet, 6 inches
Breadth over boat deck	94 feet
Height from bottom of keel to boat deck	97 feet, 4 inches
Height from bottom of keel to top of captain's house	105 feet, 7 inches

Height of funnels above casting .... 72 feet

Height of funnels above boat deck .... 81 feet, 6 inches

Distance from top of funnel to keel .... 175 feet

Number of steel decks 11

Number of watertight bulkheads 15

Passengers carried 2,500

Crew 860

Cost $10,000,000

Every line was calculated to be a little more impressive than that on any ship previously built. The great steel plates used in the hull included some as long as 36 feet, weighing 4½ tons each. Some of the great steel beams were 92 feet long, weighing 4 tons.

The rudder itself weighed 100 tons and of course was operated by electricity. The center turbine weighed 22 tons, and each of the two wing propellers 38 tons. The big boss arms from which the propellers were suspended tipped 73 tons. Even the anchor chains contributed their dimensions to the amazing total, with each link tipping 175 pounds. The 3,000,000 rivets used in construction weighed in aggregate 1,200 tons.

The White Star Line's New Triple-screw Steamers
"OLYMPIC" ☆ "TITANIC"
LARGEST AND FINEST IN THE WORLD
(SEE OVER)

A drawing comparing the length of the *Titanic* with the height of famous structures
of the world, including the Great Pyramid and the Washington Monument.

1. Washington Monument, Washington, 555 feet high
2. Metropolitan Tower, New York, 700 feet high
3. New Woolworth Building, New York, 750 feet high
4. White Star Line's Triple Screw Steamers *Olympic* and
   Titanic, 882½ feet long
5. Cologne Cathedral, Cologne, Germany, 516 feet high
6. Great Pyramid, Gizeh, Africa, 451 feet high
7. St. Peter's Church, Rome, Italy, 448 feet high

CHAPTER XXIII

GREAT MARINE DISASTERS IN RECENT YEARS

1866   January 11—Steamer *London* on its way to Melbourne, foundered in the Bay of Biscay; 220 lives lost.

1866   October 3—Steamer *Evening Star* from New York to New Orleans, foundered; 250 lives lost.

1867   October 29—Royal Mail steamers *Rhone* and *Wye*, and about 50 other vessels driven ashore and wrecked at St. Thomas, West Indies, by a hurricane; 1,000 lives lost.

1873   January 22—British steamer *Northfleet* sunk in collision off Dungeness; 300 lives lost.

1873   November 23—White Star liner *Atlantic* wrecked off Nova Scotia; 547 lives lost.

1875   May 7—Hamburg mail steamer *Schiller* wrecked in fog on Scilly Isles, 200 lives lost.

1875   November 4—American steamer *Pacific* in collision thirty miles southwest of Cape Flattery; 236 lives lost.

1878   March 24—British training ship *Eurydice*, a frigate, foundered near the Isle of Wight; 300 lives lost.

1878   September 3—British iron excursion boat *Princess Alice* sunk in collision in the Thames; 700 lives lost.

1878   December 18—French steamer *Byzantin*, sunk in collision in the Dardanelles, with the British steamer *Rinaldo*; 210 lives lost.

1880   January 31—British training ship *Atlanta* left Bermuda with 290 men and was never heard from.

1889   March 16—United States warships *Trenton*, *Vandalia*, and *Nipsic* and German ships *Adler* and *Eber* wrecked on Samoan Islands; 147 lives lost.

1891   March 17—Anchor liner *Utopia* in collision with British steamer *Anson* off Gibraltar and sunk; 574 lives lost.

1893   June 22—British battleship *Victoria* sunk in collision with the *Camperdown* off Syria; 357 lives lost.

1894   June 25—Steamer *Norge* wrecked on Rockall Reef in North Atlantic; nearly 600 lives lost.

1895   January 30—German steamer *Elbe*, sunk in collision with British steamer *Crathie* in North Sea; 335 lives lost.

1895   March 11—Spanish cruiser *Reina Regenta* foundered in Atlantic at entrance to Mediterranean; 400 lives lost.

1898   July 2—Steamship *Bourgogne* rammed British steel sailing vessel *Cromartyshire* and sank rapidly; 571 lives lost.

1904   June 15—*General Slocum*, excursion steamboat with 1,400 persons aboard; took fire while going through Hell Gate, East River; more than 1,000 lives lost.

1905   September 12—Japanese steamship *Mikasa* wrecked by explosion; 599 lives lost.

1907   February 21—English mail steamship *Berlin* wrecked off Hook of Holland; 142 lives lost.

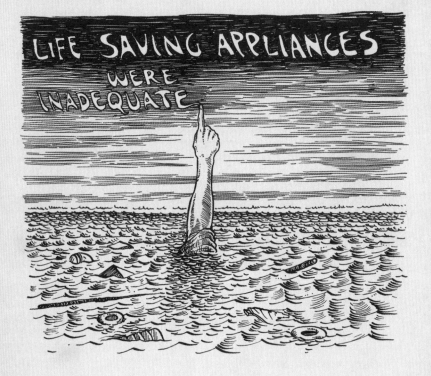

CHAPTER XXIV

THE TRAGEDY OF
THE SEA

BY REV. ANDREW JOHNSON

nd the sea gave up its dead." — Rev. 20:13

Prophets have prophesied, poets have sung of the sea, sailors have sounded its hidden depths and painters have painted its glory and its gloom.

"Roll on, thou deep and dark blue Ocean;
Ten thousand fleets sweep over thee in vain."

Today the attention of a civilized world is focused upon the fell disaster—that greatest of all disasters of the sea. The dark graveyard of the Atlantic has unfolded its bosom and taken to its trust over fifteen hundred human victims.

The catastrophe "speaks a various language" and makes a lasting impression upon art, science, business, government and religion. The startling news of the tragedy for the past days has flashed over the wires, appeared on the pages of the press and lingered on the lips of the public. It has fallen everywhere as the "words of a fatal song."

These warning tones of the *Titanic's* fate will no doubt ring loud and long in the ears of an awakened world. One of the first lessons, taught in no uncertain terms, is that of nature's supremacy over man. While man is ruler in many realms and great in his delegated lordship over many things, yet he must yield the palm, the crown and the scepter to a higher power. For with all his pomp and power and vaunted strength, with all his grand records of past achievements, he is still hedged about and hemmed in on all sides by the inflexible laws of Deity, and the stern forces of nature. What, though he has tunneled mountains, dug canyons, bridged rivers, harnessed steam, coupled together continents, captured the lightning, soared through the air as on eagle's wings, plucked messages out of the heavens and practically annihilated space, yet for all that is he not baffled and beaten by hitherto unsolved problems and unconquered forces? Like a Mohammed or a Canute, he may command the mountain to come or the waves to go, only to be defeated and disobeyed.

The refuge.

Nothing like the recent wreck in all the annals of history has so powerfully and keenly emphasized the insecurity of man and the limitations of human strength. At best, he is but a frail mortal in the midst of, and in comparison to the greater forces of nature—a mere atom, as it were, in the midst of immensity.

Relative to the famous and fateful *Titanic*, there are portrayed upon the minds of the people two pictures of sharp contrasts, the one representing strength, power and glory, the other revealing weakness, sorrow and failure. No poet's pen, no orator's tongue, no painter's brush could overdraw or overestimate the majesty of the proud mammoth ship as, launched and loaded, she starts on her maiden, her first, her last journey across the Atlantic. The inventive genius of man was taxed to its utmost in her wonderful construction and superb equipment. All the modern comforts of life, all the conveniences of land, all the luxuries of the rich, were lavished upon her. There were golf grounds, tennis courts, swimming pools, promenades, elegant parlors and concert halls—all things except a sufficient number of lifeboats in case of danger—that which should have been first and foremost was last and least in the consideration of this journey—a true type, however, of American and Anglo-Saxon life of today.

Thus fitted and furnished, the queen of the ocean, the mistress of the sea, a veritable floating palace of the deep, takes the commercial highway of the wide waters and sails for her desired port, proudly plowing the billows and breaking all records for speed. Measuring nearly nine hundred feet in length, towering like a city skyscraper, strong in her native strength and structure of steel, she poses as the very personification of safety. She claims and carries as her passengers, millionaires, bankers, world-famed editors, authors, actors, generals, pulpiteers, men of great renown and national character. She was manned and controlled by an expert and experienced sea captain and a large crew. The finest bands of music played, the sun of prosperity smiled, and it seemed that all things were replete—that nothing could be added to the comfort and convenience of those on board—all that remained for them to do was to "eat, drink and be merry" and enjoy the most pleasant journey of their lives.

Safely enfolded in the strong arms of the gigantic steel structure of the White Star line, men laughed to scorn all thoughts of danger and considered prayer for journeying mercies and providential protection needless. Why worry over wind, wave, hidden rocks and treacherous shoals; the invincible vessel is sure of her desired haven. Often when we feel we are the safest, hidden dangers lurk the nearest. So it was with the ill-fated ship *Titanic*. Sailing along under the silvery veil of a starlit night, her thousands of brilliant lights flashing out on the surrounding air, she meets a monster in her pathway. It is the crystal king of the emerald waters, the "ghostly sentinel of the banks," mantled with mist and arrayed in long robes of cloudy fog, a mountain of ice journeying southward, which claims the right of way and disputes the supremacy of the gallant ship. Then the art of man and the power of nature measured arms. The trident of Neptune was triumphant. Man's scepter fell, his crown was broken, the sullen crash of the impact of boat and berg has sounded around the world and aroused all nations. The last act of the tragedy of the *Titanic* at best can only be partially known, the full history of that final and fearful moment is buried in the two-mile tomb of the Atlantic, only to be fully revealed at the resurrection of the last day when the sea gives up its dead.

How suddenly the voice of mirth, the music of the midnight dance is changed into a doleful funeral dirge. Truly,

> "Death rides on every passing breeze,
> And lurks in every flower."

As worshippers gathered to the Lord's sanctuary on the holy Sabbath day just a week from the time of the awful disaster, they recognized, perhaps as they had not for some time, that He who walked on the storm-tossed waves of Galilee and made the yielding waters a sapphire pavement under his feet, that He who stilled the tempest with the voice of his imperative word, is the only "Sovereign of the sea," the only Master of nature.

The awful disaster brings to light more clearly than ever that the curse of the world and the crime of this age is the spirit of rivalry, the craze for speed, the desire for luxury. The train of humanity, on account of this dare-devil disposition for reckless adventure, is running so fast that it has already developed a "hot-box" and is doubtless doomed to wreck if there isn't a halt called soon. People generally are too reckless and restless. There is witnessed on every hand, in all circles and realms of twentieth century activity, an untempered and untamed mania for speed. The regular movements of modern machinery it seems can no longer satisfy this depraved and abnormal desire. Hence the strong hand of legislative enactment must, for the sake of the public welfare, put up a safeguard. Instead of luxury and speed, regard for safety and human life should and must be the rule of those who "go down to the sea in ships, that do business in great waters."

The element of heroism and self-sacrifice displayed by some of the men and the undying devotion exhibited by the wife who refused to leave her husband, are silver linings to the dark cloud of the awful disaster, are redeeming features to the dreadful calamity. This, however, is only one of the innumerable instances of the great law of vicarious sacrifice.

While death is taking such heavy toll from human life, it is well for one and all to heed the admonition, "Be ye also ready, for in such an hour as ye think not the Son of man cometh."

CHAPTER XXV

HELP FOR TITANIC SURVIVORS

THE WORLD STRAIGHTAWAY EXPRESSED ITS SYMPATHY BY OFFERING PRACTICAL HELP

he suffering survivors, on landing from the *Carpathia*, were immediately taken to hospitals and homes where they were fed, clothed, cared for and comforted and then started on their way. All over the world the people responded to the call for more lasting financial assistance and contributions were taken in the churches; funds were started by mayors and newspapers until quite a tidy sum was accumulated to help the destitute ones.

Vincent Astor, the only son of Col. John Jacob Astor, who was one of the victims, led off with his $10,000 gift to Mayor Gaynor's fund.

This contribution was delivered in the form of a check at the mayor's office by William A. Dobbyn, secretary to the late Colonel Astor, who brought it with a note from Vincent Astor.

"Will you please accept the enclosed check as a contribution from me to the fund for the needy survivors of the *Titanic* disaster?" the message ran, and Mayor Gaynor replied with this letter:

"Dear Mr. Astor,
Your generous contribution to the fund for the relief of the survivors of the *Titanic* disaster and of the dependents of those who lost their lives is at hand.

Permit me to express to Mrs. Astor and to the whole family through you my sympathy with you all in the great loss which you have sustained. My acquaintance with your father was a most agreeable one, and the oftener I met him the more his generous, superior, and democratic qualities grew on me. He was a man among men. The heroic way in which he met his death, disregarding himself and looking to the safety of others, is exactly what every one well acquainted with him knew to be the case even before authentic accounts were received.

Sincerely yours,
W. J. Gaynor, Mayor.

Vincent Astor, Esq.,
23 West Twenty-sixth Street,
New York City."

Funds were collected in all the large towns throughout the country, and contributions poured into the cities from out-of-town places. Many "benefits" were also held in the leading theaters in the cities, many theatrical stars contributing to the programs.

George M. Cohan, the actor-manager, with the assistance of friends and fellow actors, raised $9,000 for the *Titanic* sufferers. Of this amount Mr. Cohan gave individually $5,000.

By arrangement with the New York *American*, a special edition of that newspaper headed the "George M. Cohan Special" was printed and Mr. Cohan paid $5,000 for the first copy. He sold copies at the Lambs, Friars, White Rats, Comedy and Players Clubs and at the Polo Grounds. Blanche Ring paid $100 for a copy, Jerry Cohan, $100; Josephine Cohan, $100; William R.

Hearst, $200, and Mrs. Hearst, $50. The total sales, exclusive of his own copy, amounted to $1,500.

Saturday night at the Cohan Theater a special performance was given for the same purpose. The theater was crowded and about $2,500 realized. During the intermission more of the special papers were sold in the audience by Mr. Cohan and Frankie Bailey. The program was made up of stars from various theaters and vaudeville houses regardless of their syndicate and anti-syndicate affiliation.

Captain Rostron of the *Carpathia*, whose ship brought the survivors of the *Titanic*, after rescuing them, to New York.

❦ CHICAGO'S IMMEDIATE RESPONSE ❧

Although horror-stricken by the tragic details of the sinking of the *Titanic*, citizens of Chicago with the promptness which always has been characteristic of them in time of distress, arose to do their part in the nation-wide movement to provide relief for the suffering survivors, their families and the families of the dead, hundreds of whom are reported to be in want.

The appeal of Mayor Harrison confronted every citizen of Chicago, high and low, and Chicago shared generously in the big relief fund.

While leading and wealthy citizens of the city joined hands with the less fortunate in a common cause similar plans were rushed by all the churches of the city, various corporations, business houses and others.

❦ THE LONDON FUND ❧

The various London relief funds for the assistance of sufferers by the *Titanic* disaster five days after the catastrophe amounted to more than $425,000. The fund at the Mansion House alone reached $325,000. The Gaekwar of Baroda contributed $2,500, and the Prince of Wales 250 guineas ($1,250).

The Southampton fund amounted to $50,000.

Within an hour after the opening of the relief fund in Belfast $30,000 had been subscribed, including $10,000 by Lord Pirrie and $5,000 by Harland & Wolff.

When the newspapers published at great length the thrilling details of the evidence given before the senatorial commission at Washington, the extraordinary flow of money to the relief fund was proof of the wide-felt sympathy.

The total fund, including that in New York the week following the disaster, approached $1,500,000. Perhaps the most noteworthy was the *Daily Mail* fund, which was contributed exclusively by women, and amounted to $120,000. The lord mayor's fund reached $545,000 and the *Daily Telegraph's* $87,775.

The fund at Southampton amounted to $72,700, and that at Liverpool to $64,500. A large number of special performances were given at various music halls and theaters for the benefit of the sufferers.

❦─ TEMPORARY HELP ─❦

Very full of interest is the story of the relief given the survivors upon landing.

The task that was shouldered by the Women's Relief Committee of supplying some of the immediate needs of the *Titanic's* survivors took long forward strides Saturday, the day after the *Carpathia* came in, so that the corridors and wards of St. Vincent's Hospital were astir with the distribution of warm clothes. Before nightfall many of the shipwrecked were moving onto their destinations.

It was the idea of the committee of women, organized on Tuesday evening by Mrs. Nelson H. Henry, wife of the Surveyor of the Port, that there should be hands extended to the people and particularly women's hands when the *Carpathia* came in, but the relief they offered was only for immediate needs, and the larger fund collected by Mayor Gaynor and others was for the work of more permanent benevolence for those who lost so much when the big ship sank.

One Friday evening, it was announced that the committee had received plenty for all the work that it could do, but this had to be repeated, for the morning's mail brought in a flood of contributions to the amount of more than $1,700. Four of the benefit performances offered were accepted, but the committee asked all others to extend the offers to the Red Cross as a contribution from the Women's Relief Committee.

The committee's work in its rooms on the sixth floor of the Metropolitan Life Building was divided into two departments. One took care of the receiving and distributing of clothes, and the other was devoted to the questions of immediate relief, of money, transportation, temporary homes, and arrangements for employment later.

All this was rapidly reduced to a catalogue, so that when word came from St. Vincent's or other hospitals and homes where survivors were taken, accompanied by the certificate of assent from the Commissioners of Immigration, the committee knew just what was wanted, just what size clothes, just where the people wanted to go, and just how much money was necessary. The committee offered help for the first four weeks after the shipwreck. Clothing, a railroad ticket, perhaps, and a little money was bestowed in each case along with a deal of comforting.

◈◄ PROMINENT WOMEN WORKED HARD ►◈

But all this was no simple undertaking, and the offices were jammed all the time. Women of prominence could be seen moving about from task to task. Miss Anne Morgan was always busy, Mrs. August Belmont and Mrs. Eugene Kelly helped with this case and that, Mrs. Edward Hewitt was a tower of strength, and Mrs. Henry Dimock was tireless as her bundles of clothing arrived, stack on stack, and her motor car carried her from one errand to another.

Representatives from different houses that had opened their doors to survivors would appear with the names and conditions of those who were ready to move on. A priest from the Swedish Home was there to arrange for clothing and money for thirteen charges. A big man from the Salvation Army arrived with the list of those under his care. He had been down to arrange for the transportation of Mrs. Emily Goldsmith and her little son, who must move on to Detroit without the husband and father that sailed with them from Southampton.

There was one moment when the women paused to shake their heads sadly, for an application had come in for an outfit for a young girl who lost her brother in the wreck. And the man who brought the requisition asked that the dresses be not black for the girl would not give up hope.

Of the 106 *Titanic* people who were taken to St. Vincent's Hospital, fully forty resumed their journey Saturday. A dazed girl sailed back to Finland on Wednesday. Her brother, her uncle, and the man she was to marry were lost. A slender little Swedish woman hovered over her two babies, patting their hair and smoothing down the new dresses that came from the committee of women. She had one terrible moment when she started down the rope to the already lowered lifeboat and knew that she could carry only the smaller child. The three-year-old girl she could not carry, but the little girl clung terrified to the mother's skirt and did not release the hold till all three were in the lifeboat.

CHAPTER XXVI

SOME PATHETIC FEATURES OF THE TRAGEDY

SO MANY PEOPLE NEAR SAFETY JUST MISSED IT — HELPLESS ONES LEFT AND THEIR PROTECTORS TAKEN

itiful tales were related by some of the steerage passengers of the *Titanic* as they came off the *Carpathia*. Few of the passengers were met by relatives or friends and a majority were taken in charge by charitable persons.

A pathetic incident of the steerage was the placing of seven children—four girls and three boys—into one of the lifeboats. Their parents were lost. Two of the little ones, whose names could not be ascertained, were taken to hospitals. One had scarlet fever and the other meningitis.

◄ SOME DIED IN LIFEBOATS ►

H. Haven, of Indianapolis, said the *Titanic* was going at high speed when she struck and that the helmsman apparently had seen danger and put the helm over, for the boat veered to port and struck the iceberg a glancing blow. This ripped off a large section of the plates on the starboard side and the water began to pour in. He said:

"There was a great rush for the lifeboats as soon as it was known that there was any real danger. So precipitate was this rush that many in apparent frenzy jumped over the ship's railing into the sea. A remarkable thing was that the lights continued to burn, although the *Titanic* settled lower and lower.

"When we were at some distance from the sinking ship and could still see the figures of hundreds of people on deck at the railings there were several explosions in the ship. More people went overboard. Presently the *Titanic* buckled amidships, and we could see the people sliding off into the water, both fore and aft. Then the boat settled somewhat by the bow, the lights went out and that was the last we saw of the *Titanic*.

"The temperature must have been below freezing, and neither the men nor the women in my boat were warmly clad. Several of them died. The officer in charge of the lifeboat decided it was better to bury the bodies. So they were weighted and put overboard. We could also see similar burials taking place from other lifeboats that were all around us.

"Of course at that time we did not know the *Carpathia* was near."

❖― HUNDREDS DROWNED ―❖

August Wennerstrom, a Swede, spied a collapsible boat behind one of the smokestacks as the vessel was sinking. With three other men he managed to tear it from its lashings and the four jumped overboard with it. The boat overturned four times, but each time they managed to right it and finally all of them were saved by the *Carpathia*. While drifting about, Wennerstrom said he saw at least two hundred men in the water who were drowned.

People await the arrival of the *Titanic* survivors.

⟫⟨ CREW OBLIGED TO JUMP ⟩⟫

The chief steward of the *Carpathia* explained the large number of the crew saved by saying that the majority of them had jumped from the *Titanic* and were picked up by the boats.

⟫⟨ SORROW INSTEAD OF SURPRISE ⟩⟫

This message was received from London two days after the *Carpathia* came in by James W. Van Billiard, of North Wales, Pa.: "Austin and two oldest children sailed on *Titanic*. Maude."

It explained that Austin Van Billiard, son of James W. Van Billiard, Burgess of North Wales and a wealthy marble dealer, accompanied by his two eldest children, James, aged eleven, and Walter, aged nine, had sailed from Liverpool on the *Titanic*.

It further explained to the Van Billiard family that it was their son whose name appeared in the list of steerage passengers who went down, and not some one with a similar name as they had believed.

⠀⠀MEN HUNG ON RAFTS⠀⠀

One version of the deaths of John Jacob Astor and William T. Stead was told by Philip Mock, who with his sister, Mrs. Paul Schabert, were among the survivors.

"Many men were hanging on to the rafts in the sea," said Mr. Mock. "William T. Stead, the author, and Col. John Jacob Astor clung to a raft. Their feet became frozen and they were compelled to release their hold. Both were drowned."

⠀⠀ALL THAT IS LEFT⠀⠀

In the children's ward at St. Vincent's was a little girl four years old who was brought off the sinking *Titanic*. Her name, she thought, was Annie Karens. She lisped it and wanted her father and mother. People kept telling the child that Mamma might come after a while and that Papa might come, too.

The wives and relatives and friends of the crew of the *Titanic* gathered in the early hours of the morning of the 19th at the White Star offices in Southampton, England, to wait for the list of those officers and men who had been saved. In some cases the posting of the list brought relief, but the majority went away with their worst fears confirmed.

CHAPTER XXVII

SOME FORTUNATE CIRCUMSTANCES

IN THE MIDST OF THE GLOOM OF THE TRAGEDY THERE ARE STILL SOME BRIGHT SPOTS — BETTER HIMSELF THAN HIS WARDROBE

lfred von Drachstedt, a tall, blonde German youth of twenty years, who says that he has the right to prefix "baron" to his name, appeared Thursday night on the *Carpathia* attired in a sweater, a pair of trousers, and a life preserver and with only a few German marks in his pocket. He left on the *Titanic* 750 German marks and a wardrobe.

It was an elaborate wardrobe that young Von Drachstedt left behind him, and he felt bad over its loss, though admitting that he was glad to have arrived himself.

To begin with, it was a brand new wardrobe, and it cost, according to his itemized account, just $2,133, counting in the jewelry, walking sticks, two sets of toilet articles, and a fountain pen that went with it.

The young man lives in Cologne and his mother is a widow. It was his first trip from home.

❧ SAVED BY DOING HIS DUTY ❧

Rev. James M. Gray, dean of the Moody Bible Institute, probably owed his life to his conscientious desire to return to America in time to preach the baccalaureate sermon to the graduating class of the institute. He was about to start for home when Rev. Dr. Harold urged him to remain and embark on the *Titanic* on her maiden voyage. He refused to do so on the plea that he must be in Chicago to preach to the graduates. He took another steamship a week earlier.

❈⤙ A POST CARD PROPHECY ⤚❈

A picture postal card, with the following jingle, bore the first news to Rev. and Mrs. Mawbrey E. Collett, of Port Byron, New York, that their son, the Rev. Sidney C. Stuart Collett, had embarked on the ill-fated *Titanic*. The card, bearing a picture of the *Titanic*, said:

> "Mother put the kettle on, let's have a cup of tea
> Ready for the dear old 'Sid,' who's coming home from sea;
> You'll be glad to see him, and kiss him with delight,
> So mother put the kettle on, I'm coming home all right.
> (Signed) Sid."

That was all the news of the young traveler that they had until they read his name among the list of survivors.

❈⤙ TRIBUTE TO MARCONI ⤚❈

Oscar Straus, brother of Isidor Straus, the great philanthropist who lost his life on the *Titanic*, paid a high tribute to the genius of Marconi and said that he hoped a monument may be erected to the inventor during his lifetime.

> "But for the genius of Marconi, every soul on the *Titanic* would probably have been drowned and we would not have known what happened. To him the survivors owe

their lives, and no tribute we can pay would be too great.

"What he has done to safeguard the lives of those who travel on the seas should not be underestimated, and his inventions have made him one of the great figures in the world today. I should like to see a monument erected to him while he lives so that he may see that the world appreciates what he has done for humanity."

Survivors of the *Titanic* disaster at Millbay Docks in Plymouth, May 1, 1912.

⟪ ESCAPED ON ICE ⟫

A huge cake of ice was the means of aiding Emile Portaluppi, of Aricgabo, Italy, in escaping death when the *Titanic* went down. Portaluppi, a second-class passenger, was awakened by the explosion of one of the boilers of the ship. He hurried to the deck, strapped a life preserver around him and leaped into the sea.

With the aid of the preserver and by holding to a cake of ice he managed to keep afloat until one of the lifeboats picked him up. There were thirty-five other people in the boat when he was hauled aboard.

CHAPTER XXVIII

VARIOUS DESCRIPTIONS OF HOW THE TITANIC DISAPPEARED

EVERY SURVIVOR WAS LEFT WITH A VIVID IMPRESSION OF THE SHIP'S TRAGEDY — END OF TITANIC APPALLING

ne of the most stirring accounts of the wreck and its after effects was told by William Smith, assistant manager for L. E. Waterman, 115 South Clark Street, Chicago.

With tears gushing from his eyes, though he tried to wink them away, Smith told of the thrilling experiences of Mrs. Harry Collyer of Bishopstoke, near Southampton, England, and her eight-year-old daughter Marjorie.

The Collyers numbered the husband, Harry Collyer, thirty years old; his wife and daughter. They had booked passage on the steamer New York, which was delayed because of the British coal strike, and they were transferred to the *Titanic*.

Collyer, who perished on the *Titanic*, had purchased through tickets for the family to Payette, Idaho, where he intended to buy a half-interest in a ten-acre apple orchard. The Collyers had sold their little grocery at Bishopstoke, and the husband had all his money and valuables sewed up in his clothes. Mr. Smith told Mrs. Collyer's story:

❦ WOMAN DESCRIBED DISASTER ❦

"Mrs. Collyer told me a terrible story of the disaster.

It was bad enough to meet her at the dock when the *Carpathia* came in. I would not suffer that experience again for $1,000.

"When the *Titanic* struck the iceberg, the Collyers were awakened from slumber in their berths and rushed to the deck, thinly clad. Some one called out that all the passengers should put on life preservers. Collyer rushed away to find three of them for his family and himself.

"His wife never saw him again. She was thrown into a lifeboat with other women. Just before this she said she saw three lifeboats, one after another, overturned with their human freight. It was this that frightened the women on board and made them reluctant to enter the boats.

"The result was, Mrs. Collyer said, that women were torn from the arms of their loved ones and thrown bodily into the life craft.

"Officers stood by with pistols to keep away the men from the steerage, who on at least one occasion attempted a rush. When occasion warranted the officers did not scruple to fire. One of the men from the steerage jumped into one of the lifeboats. The officer in charge threw him out into the water to drown.

"Mrs. Collyer thought that the disaster caught the crew of the *Titanic* unawares, as she said there was not a proper response when the call to the lifeboats was issued.

"The end of the *Titanic* she described as appalling, as seen from the lifeboats through the starlit night. First one end of the steamer lifted, then the other; then, with a great wail from hundreds still on board, it sank.

Survivors of the *Titanic*, in one of her collapsible lifeboats.

"For one hour, she averred, the screams continued, right up to the time when the *Titanic* disappeared forever, and she said that this was the unforgettable impression of the wreck for her. The lifeboats had all they could do to preserve their equilibrium and to prevent collision with the icebergs.

"The Collyers are absolutely destitute, as the husband carried to the bottom with him all they had in the world."

❖⊷ NO SEARCHLIGHT ⊶❖

Miss Constance Willard, of Duluth, Minn., who left the *Titanic* twenty minutes before the vessel sank, recounted an interesting experience.

"One subject talked of after we were on board the *Carpathia* was the fact the *Titanic* had no searchlight. The crew said that it had been the intention of the owners to equip the vessel with a searchlight after its arrival in New York.

"When I reached the deck after the collision the crew were getting the boats ready to lower, and many of the women were running about looking for their husbands and children. The women were being placed in the boats, and two men took hold of me and almost pushed me into a boat. I did not appreciate the danger and I struggled until they released me.

"'Do not waste time; let her go if she will not get in,' an officer said. I hurried back to my cabin again and went from cabin to cabin looking for my friends, but could not find them. A little English girl about fifteen years old ran up to me and threw her arms about me.

≪⊷ Hurried Aboard a Boat ⊶≫

"'O, I am all alone,' she sobbed, 'won't you let me go with you?' I then began to realize the real danger and saw that all but two of the boats had been lowered. Some men called to us and we hurried to where they were loading a boat. All the women had been provided with life belts. As the men lifted us into the boat they smiled at us and told us to be brave. The night was cold and the men who were standing about, especially the steerage passengers, looked chilled, but the men who were helping the women into the boats seemed different. Even while they smiled at us, great beads of perspiration stood out on their foreheads.

≪⊷ Begged Her to Take Child ⊶≫

"I never will forget an incident that occurred just as we were about to be lowered into the water. I had just been lifted into the boat and was still standing, when a foreigner rushed up to the side of the vessel and holding out a bundle in his arms cried with tears running down his face:

"'O, please, kind lady, won't you save my little girl, my baby. For myself it is no difference, but please, please take the little one.' Of course, I took the child. Most women were compelled to stand in the boats because they all wore the life belts, which made it almost impossible to sit down.

"In our boat there were seven men, about twenty women, and several children. The night was dark. Twenty minutes after leaving the *Titanic* we heard an explosion and the vessel appeared to split in two and sank. Then a foreign woman in our boat began singing a hymn, and we all joined, although few knew the words. All around us we heard crying and sobbing for perhaps three minutes."

❖ NO IDEA BOAT WOULD SINK ❖

John B. Thayer, Jr., whose father, the second vice-president of the Pennsylvania Railroad Company, went down with the *Titanic*, while his mother was saved, dictated at his home in Haverford, in the presence of members of his family and officers of the railroad company, an account of his thrilling experience in the great sea tragedy. Young Thayer, who is seventeen years old, said in part:

"Father was in bed and Mother and myself were about to get into bed. There was no great shock. I put on an overcoat and rushed up on 'A' deck on the port side, but saw nothing there. I then went down to our room and my father and mother came on deck with me. The ship had then a fair list to port.

❖ Described Farewell to Mother ❖

"We then went down to our rooms, all dressing quickly. We all put on life preservers, and over these we put our overcoats. Then we hurried up on deck and walked around until the women were all ordered to collect on the port side. Father and I said goodbye to Mother at the top of the stairs on the A deck.

"As at this time we had no idea the boat would sink, we walked around. We met the chief steward of the main dining saloon and he told us that Mother had not yet taken a boat, and he took us to her.

"Father and Mother went ahead, and I followed. A crowd got in front of me and I was not able to catch them and lost sight of them. That is the last time I saw my father. This was about half an hour before the ship sank. I then went to the starboard side with Milton C. Long, of New York.

"On the starboard side the boats were getting away quickly. We thought of getting into one of the boats, but there seemed to be such a crowd around I thought it wouldn't do to make any attempt.

❦ Jumped into Ocean; Found Boat ❦

"About this time people began jumping from the stern. I thought of jumping myself, but was afraid of being stunned on hitting the water. As the boat started to sink we stood by the rail. Long and myself said good-bye to each other and jumped up on the rail. He did not jump clear, but slid down the side of the ship. I never saw him again.

"I jumped out feet first, went down, and as I came up I was pushed away from the ship by some force.

"I was sucked down again and as I came up I was pushed out again and twisted around by a large wave coming up in the midst of a great deal of small wreckage. My hand touched the cork fender of an overturned lifeboat. I looked up and saw some men on the top. One of them helped me up. In a short time the bottom was covered with about twenty-five or thirty men.

❦⤚ Rescue Boat Arrived ⤙❧

"The assistant wireless operator was right next to me, holding on to me and kneeling in the water. We all sang a hymn and said the Lord's Prayer, and then waited for morning to come. The wireless man raised our hopes by telling us that the *Carpathia* would be up in about three hours. About 3:30 or 4 o'clock some men on our boat on the bow sighted its mast lights.

"Two boats from the *Carpathia* came up. The first took half and the other took the balance, including myself. In about a half or three-quarters of an hour later we were picked up by the *Carpathia*."

CHAPTER XXIX

U.S. SENATORS OBTAIN FACTS OF WRECK

SPEED OF SHIP NOT LESSENED ON WARNING —
WITNESSES ALSO SHOWED LACK OF SMALL BOATS
COST MANY LIVES — ISMAY DESCRIBED WRECK —
DENIED HE FLED BEFORE WOMEN HAD CHANCE TO
LEAVE THE VESSEL — DESCRIBED RESCUE EFFORTS

he seriousness of the inquiry by the United States Senate investigating committee into the *Titanic* disaster was disclosed when Senator William Alden Smith of Michigan, the chairman, at first flatly refused to let any of the officers or the two hundred-odd members of the crew of the sunken steamship get beyond the jurisdiction of the United States government. The men were all to have sailed back home on the steamer *Lapland*.

Later it was decided that the greater part of the crew would be permitted to sail, but that the twelve men and four officers among the survivors under subpoena, together with J. Bruce Ismay, would not be allowed to depart.

It was explained that Mr. Ismay was anxious to leave at once for Europe, as he had been worn out by his experiences, and felt the need of returning quickly to his English home for a rest. His pleas, however, were unavailing.

❦ MEN WHO TESTIFIED ❦

The first day brought out important features in connection with the wreck. These were disclosed in the examination of Mr. Ismay, Arthur Henry Rostron, captain of the rescue ship *Carpathia*, and Second Officer Lightoller of the *Titanic*, William Marconi, inventor of the wireless telegraph; Thomas Cottam, the wireless operator of the *Carpathia*, and others.

Among other things, the first day's testimony showed: That the biggest ship ever built sank in midocean because it was being rushed forward almost at top speed and crashed into a field of icebergs after warnings had been given to look out. That the small number of lives saved was due to the fact there were not enough lifeboats on board to accommodate the passengers.

ISMAY DESCRIBED THE WRECK

Because of his position as managing director of the White Star Line, the testimony of Mr. Ismay was the most important given.

Mr. Ismay, who plainly showed his nervousness while on the stand, told in whispers of his escape from the sinking liner from the time he pushed away in a boat with the women until he found himself, clad in his pajamas, aboard the *Carpathia*.

He was not sure in just what boat he left the *Titanic*, nor was he sure how long he remained on the liner after it struck. He added, however, that before he entered a lifeboat he had been told that there were no more women on the deck.

Mr. Ismay denied that there had been any censoring of messages from the *Carpathia*. Other witnesses bore him out in this, with the explanation that the lone wireless operator on the rescue ship was unable to send matter for the press.

TEXT OF ISMAY TESTIMONY

Mr. Ismay, in response to Senator Smith's questionings, gave an account of his experiences.

"As near as I remember, it was the 1st of April that the *Titanic* made its trial trip, which was perfectly satisfactory. On the voyage over, we left Southampton at 12 o'clock and arrived at Cherbourg that evening, having made the run at sixty-eight revolutions. We left Cherbourg and proceeded to Queenstown, arriving there, I think, at midday on Thursday. We ranged, I think, about seventy revolutions. We embarked passengers and proceeded at seventy revolutions. I am not absolutely clear on the run on the first day. I think it was between 464 and 474 miles. The second day we proceeded at seventy-two revolutions, the third day at seventy-five. I think that day we ran either 576 or 579 miles. The weather continued fine, except for about ten minutes of fog one evening. The accident took place on Sunday night. The exact time I don't know. I was in bed asleep when it happened. The ship sank, I am told, at 2:20 in the morning. The ship had never been at full speed. That would have been seventy-eight revolutions, working up to eighty. It hadn't all its boilers on. I may say that it was intended, if we had fair weather Monday afternoon or Tuesday, to drive the steamship at full speed. Unfortunately the catastrophe prevented this.

"I presume the impact awakened me. I lay for a minute or two and then I got up and went into the passageway, where I met a steward and asked him what was the matter. He replied, 'I don't know, sir.' Then I went back to my stateroom, put on my overcoat and went up to the bridge, where I saw Captain Smith. 'What has happened?' I asked him. 'We have struck ice,' he replied. I asked if the injury was serious, and he said he thought so. Then I came down and in an entryway saw the chief engineer. I asked him if he thought there was any serious injury. He said he believed there was. Walking along the deck I met an officer on the starboard side and assisted him as best I could in getting out the women and children. I stayed up on deck until the starboard collapsible boat was lowered."

Mr. Ismay stated that an official representative of the builders, Mr. Thomas Andrews, was on board to see that everything was satisfactory and wherein improvements might be made, but he was lost.

"Did you or the captain ever consult about movement of the ship?"

"Never."

❦– Slow Increase in Speed –❦

"Was it supposed that you could reach New York by 5 o'clock Wednesday morning without putting the steamship to its full capacity?"

"Oh, yes. Nothing was to be gained by arriving sooner than that."

Mr. Ismay testified that the revolutions were being gradually increased, as was customary with a new ship. The speed on Saturday was 75 revolutions, but that was nothing to full speed. Mr. Ismay did not know ice had been reported, and had never seen an iceberg. He expected that some time Sunday night they would come into the ice region.

"Did you have any consultation with the captain regarding this matter?"

"Absolutely none. It was entirely out of my province. I was simply a passenger aboard the ship."

"On which decks were the boats?"

"The lifeboats were all on one deck—the sun deck."

❰❰ Four Men in His Boat ❱❱

They were filled, a crew put in, and they were sent away. There were four men aboard the boat on which Mr. Ismay escaped.

J. Bruce Ismay, one of the *Titanic* survivors,
testifies at the U.S. Senate inquiry into the disaster.

⟨⟨ Women Sent Away First ⟩⟩

Mr. Ismay could not say that all the women and children had been taken off. In his boat there were about forty-five people. Three other boats he saw were loaded about the same. There was no struggle by men to get into the boats and the women were taken just as they came. Mr. Ismay said he was on the *Titanic* practically until it sank, perhaps an hour and a quarter.

"What were the circumstances of your departure from the ship?" asked Senator Smith.

"I was immediately opposite the lifeboat. A certain number of people were in it. An officer called to know if there were any more women. There were no women in sight on the deck then. There were no passengers about and I got in."

Nearly all the passengers Mr. Ismay saw had on life preservers. He did not see anyone jump into the sea. They steered their lifeboats toward a distant light and spent about four hours in the open sea.

"How many lifeboats were there?"

"Twenty altogether, I think; sixteen of them wooden lifeboats, but I am not absolutely certain."

Mr. Ismay said the sea was very calm.

"What can you say about the sinking and disappearance of the ship?" asked Senator Smith.

"Nothing; I did not see it go down."

"I was sitting with my back to the ship; I did not wish to see it go. I was pushing with an oar. I am glad I did not see it."

⟨⟨ CONFORMED TO BOARD RULES ⟩⟩

Mr. Ismay said the *Titanic* conformed to the British Board of Trade's requirements, else it could not have sailed. The lifeboats were the *Titanic*'s own and not borrowed from any other ship of the White Star line. Mr. Ismay had nothing to do with the selection of the men in his own lifeboat; they were designated by Mr. Wild, the chief officer.

⫷ SAFER THAN OTHER SHIPS ⫸

Senator Smith wished to know how much water the ship could hold without sinking.

"The ship was especially constructed so as to float with any two compartments—any two of the biggest compartments—full of water, and I think I am right in saying there are few ships today of which the same can be said. When we built the ship we had this in mind. If the ship had hit the ice head-on, in all human probability that ship would have been afloat today, but the information I received is that it struck a glancing blow between the end of the forecastle and the captain's bridge."

Mr. Ismay feared all the women and children were not saved. He could say nothing of equipment and so on, except that the Board of Trade rules had been complied with and that all data and information was at the committee's disposal. He had made no attempt to interfere with the wireless service in any way.

⫷ TESTIMONY OF CAPTAIN ROSTRON ⫸

Captain Rostron of the *Carpathia* followed Mr. Ismay. He told Mr. Smith that he had been captain of the *Carpathia* since last January, but that he had been a seaman twenty-seven years.

"What day did you last sail from New York with the *Carpathia*?" asked Senator Smith.

"April 11," said Captain Rostron, "bound for Gibraltar."

"How many passengers did you have?"

"I think 120 first-class, 50 second-class, and about 565 third-class passengers."

Wireless anarchy.

"Tell the committee all that happened after you left New York."

"We backed out of the dock at noon, Thursday. Up to Sunday midnight we had fine, clear weather. At 12:35 Monday morning I was informed of the urgent distress signal from the *Titanic*."

"By whom?"

"The wireless operator and first officer. The message was that the *Titanic* was in immediate danger. I gave the order to turn the ship around. I set a course to pick up the *Titanic*, which was fifty-eight miles west of my position. I sent for the chief engineer; told him to put on another watch of stokers and make all speed for the *Titanic*. I told the first officer to stop all deck work, get out the lifeboats, and be ready for any emergency. The chief steward and doctors of the *Carpathia* I called to my office and instructed as to their duties. They were instructed to be ready with all supplies necessary for any emergency."

❦ HOW SURVIVORS WERE FOUND ❦

Arriving on the scene of the accident, Captain Rostron testified, he saw an iceberg straight ahead of him, and, stopping at 4 A.M., he picked up the first lifeboat.

"By the time I got the boat aboard day was breaking," said the captain. "In a radius of four miles I saw all the other lifeboats. On all sides of us were icebergs; some twenty were 150 to 200 feet high, and numerous small icebergs, or 'growlers.' Wreckage was strewn about us. At 8:30 all the *Titanic*'s survivors were aboard."

❦ The Service of Prayer ❦

Then, with tears filling his eyes, Captain Rostron said he called the purser.

"I told him," said Captain Rostron, "I wanted to hold a service of prayer—thanksgiving for the living and a funeral service for the dead. I went to Mr. Ismay. He told me to take full charge. An Episcopal clergyman was found among the passengers and he conducted the services."

Three members of the *Titanic*'s crew were taken from the lifeboats, dead from exposure. They were buried at sea.

Asked about the lifeboats, Captain Rostron said he found one among the wreckage in the sea. The lifeboats on the *Titanic*, Captain Rostron said, were all new and in accordance with the British regulations.

"Was the *Titanic* on the right course when it first spoke to you?" Senator Smith asked.

"Absolutely on its regular course bound for New York," said the captain. "It was in what we call the southerly to avoid icebergs."

❦ Silent as to Warning ❧

Captain Rostron declined to say if Captain Smith had warning enough and might have avoided the ice if he had heeded.

"Would you regard the course taken by the *Titanic* in this trial trip as appropriate, safe and wise at this time of the year?" Senator Smith asked.

"Quite so."

"What would be safe, reasonable speed for a ship of that size and in that course?"

"I didn't know the ship," the captain said, "and therefore cannot tell. I had seen no ice before the *Titanic* signaled us, but I knew from its message that there was ice to be encountered. But the *Carpathia* went full speed ahead. I had extra officers on watch and some others volunteered to watch ahead throughout the trip."

❦ *Carpathia* Had Twenty Lifeboats ❧

Captain Rostron said the *Carpathia* had twenty lifeboats of its own, in accordance with the British regulations.

"Wouldn't that indicate that the regulations are out of date, your ship being much smaller than the *Titanic*, which also carried twenty lifeboats?" Senator Smith asked.

"No. The *Titanic* was supposed to be a lifeboat itself."

"You say that the captain of a ship has absolute control over the movements of his vessel?"

"Yes, by law that is the rule," Captain Rostron answered. "But suppose we get orders from the owners of our ship to do a certain thing. If we do not execute that order we are liable to dismissal. When I turned back for New York with the rescued I sent a message to the Cunard line office stating that I was proceeding to New York unless otherwise ordered. I then immediately proceeded. I received no order to change my course."

Senator Smith said some complaint had been heard that the *Carpathia* had not answered President Taft's inquiry for Major Butt. Captain Rostron declared a reply was sent "not on board."

❦ Caught Appeal by Chance ❧

Absolutely no censorship was exercised, he said. The wireless continued working all the way in, the Marconi operator being constantly at the key. In discussing the strength of the *Carpathia*'s wireless, Captain Rostron said the *Carpathia* was only fifty-eight miles from the *Titanic* when the call for help came.

"Our wireless operator was not on duty," said Captain Rostron, "but as he was undressing he had his apparatus to his ear. Ten minutes later he would have been in bed and we never would have heard."

⊰ MARCONI ON THE STAND ⊱

William Marconi, the wireless inventor, took the stand as soon as the hearing was resumed. He said he was the chairman of the British Marconi Company. Under instructions of the company, he said, operators must take their orders from the captain of the ship on which they are employed.

"Do the regulations prescribe whether one or two operators should be aboard the ocean vessel?"

"Yes, on ships like the *Titanic* and *Olympic*, two are carried," said Mr. Marconi. "The *Carpathia*, a smaller boat, carries one. The *Carpathia* wireless apparatus is a short distance equipment. The maximum efficiency of the *Carpathia* wireless, I should say, was 200 miles. The wireless equipment on the *Titanic* was available 500 miles during the daytime and 1,000 miles at night."

"Do you consider that the *Titanic* was equipped with the latest improved wireless apparatus?"

"Yes; I should say that it had the best."

Senator Smith asked if amateur or rival concerns interfered with the wireless communication of the *Carpathia*.

⊰ INTERFERENCE BY OUTSIDERS ⊱

"I am unable to say. Near New York I have an impression there was some slight interference, but when the *Carpathia* was farther out in touch with New York and Nova Scotia there was practically no interference."

"Did you hear the captain of the *Carpathia* say in his testimony that they caught this distress message from the *Titanic* almost providentially?" asked Senator Smith.

"Yes, I did. It was absolutely providential."

"Ought it not be incumbent upon ships to have an operator always at the key?"

"Yes, but the ship owners do not like to carry two operators when they can get along with one. The smaller boat owners do not like the expense of two operators."

❮ TESTIMONY OF SECOND OFFICER ❯

Charles Herbert Lightoller, second officer of the *Titanic*, said he understood the maximum speed of the *Titanic*, as shown by its trial tests, to have been 22½ to 23 knots.

Senator Smith asked if the rule requiring life-saving apparatus to be in each room for each passenger was complied with.

"Everything was complete," said Lightoller. During the tests, he said, Captain Clark of the British Board of Trade was aboard the *Titanic* to inspect its life-saving equipment.

"How thorough are these captains of the Board of Trade in inspecting ships?" asks Senator Smith.

"Captain Clark is so thorough that we called him a nuisance."

Lightoller said he was in the sea with a life belt on one hour and a half after the *Titanic* sank. When it sank he was in the officers' quarters and all but one of the lifeboats were gone. This one was caught in the tackle and they were trying to free it.

❮ Had Warning of Icebergs ❯

Lightoller said that on Sunday he saw a message from "some ship" about an iceberg ahead. He did not know the *Amerika* sent the message, he testified.

The ship was making about 21 to 21½ knots, the weather

was clear and fair, and no anxiety about ice was felt, so no extra lookouts were put on.

"When Capt. Smith came on the bridge at five minutes of 9, what was said?"

"We talked together generally for twenty or twenty-five minutes about when we might expect to get to the ice fields. He left the bridge, I think, about twenty-five minutes after 9 o'clock, and during our talk her told me to keep the ship on its course, but that if I was the slightest degree doubtful as conditions developed to let him know at once."

"What time did you leave the bridge?"

"I turned over the watch to First Officer Murdock at 10 o'clock. We talked about the ice that we had heard was afloat, and I remember we agreed we should reach the reported longitude of the ice floes about 11 o'clock, an hour later. At the time the weather was calm and clear. I remember we talked about the distance we could see. We could see stars in the horizon. It was very clear."

Lightoller testified that the *Titanic's* decks were absolutely intact when it went down. The last order he heard the captain give was to lower the boats.

The last boat, a flat collapsible, to put off was the one on top the officers' quarters. Me jumped upon it on deck and waited for the water to float it off. Once at sea it upset. The forward funnel fell into the water, just missing the raft, and overturning it. The funnel probably killed persons in the water.

"This was the boat I eventually got on. No one was on it when I reached it. Later about thirty men clambered out of the water on to it. All had on life preservers."

"Did any passengers get on?" asked Senator Smith.

"J. B. Thayer, Colonel Gracie and the second Marconi operator were among them. All the rest taken out of the water were firemen. Two of these died that night and slipped off into the water. I think the senior Marconi operator was one of the three. We took on board all we could and there were no others in the water near at hand."

When Lightoller left he saw no women or children on board, though there were a number of passengers on the boat deck. The passengers were selected to fill the boat by sex, Lightoller himself putting on all the women he saw, except the stewardesses. He saw some women refuse to go.

❊⊷ Twenty-Five in First Boat ⊶❊

In the first boat to be put off Lightoller said he put twenty to twenty-five. Two seamen were placed in it. The officer said he could spare no more, and that the fact that women rowed did not show the boat was not fully equipped.

At that time he did not believe the danger was great. Two seamen placed in the boat, he said, were selected by him, but he could not recall who they were. He said he named them because they were standing near. The second boat carried thirty passengers, with two men.

"By the time I came to the third boat I began to realize that the situation was serious, and I began to take chances. I filled it up as full as I dared, sir—about thirty-five, I think."

❊⊷ Ran Short of Seamen ⊶❊

In loading the fourth lifeboat, Lightoller said he was running short of seamen.

"I put two seamen in and one jumped out. That was the first boat I had to put a man passenger in. He was standing nearby and said he would go if I needed him.

"I said, 'Are you a sailor?' and he replied that he was a yachtsman. Then I told him that if he was sailor enough to get out over the bulwarks to the lifeboat, to go ahead. He did, and proved himself afterward to be a brave man. I didn't know him then, but afterward I looked him up. He was Major Peuchen of Toronto."

On the fifth boat Lightoller had no particular recollection.

"The last boat I put out, my sixth boat," he said, "we had difficulty finding women. I called for women and none were on deck. The men began to get in—and then women appeared. As rapidly as they did, the men passengers got out of the boat again."

"The boat's deck was only ten feet from the water when I lowered the sixth boat. When we lowered the first the distance to the water was seventy feet."

All told, Lightoller testified, 210 members of the crew were saved.

"If the same course was pursued on the starboard side as you pursued on the port filling boats, how do you account for so many members of the crew being saved?" asked Chairman Smith.

"I have inquired especially and have found that for every six persons picked up five were either firemen or stewards."

Some lifeboats, the witness said, went back after the *Titanic* sank and picked up men from the sea.

Lightoller said he stood on top of the officers' quarters and as the ship dived he faced forward and dived also.

"I was sucked against a blower and held there. A terrific gust came up the blower—the boilers must have exploded—and I was blown clear—barely clear. I was sucked down again, this time on the 'Fidley' grating."

Colonel Gracie's experience was similar. Lightoller did not know how he got loose, perhaps another explosion. He came up by a boat, on which he clambered.

◄ TESTIMONY BY RELIEF MAN ►

Thomas Cottam, aged twenty-one, of Liverpool, the Marconi operator on the *Carpathia*, was the next witness.

He said he had no regular hours for labor on the *Carpathia*. Previous witnesses had testified he was not "on duty" when he received the *Titanic*'s signal for help. He was uncertain whether he was required to work at night. He had not closed his station for the night, which is accomplished by switching the storage battery out. He was listening for a confirmation message from the *Parisian*, while he was preparing to retire, and caught the *Titanic*'s distress signal by chance.

"When you got the distress message from the *Titanic* Sunday night, how did you get it?"

"I called the *Titanic* myself, sir."

"Who told you to call the *Titanic*?"

"No one, sir; I did it of my own free will."

"What was the answer?"

"'Come at once,' was the message, sir."

"I was in communication with the *Titanic* at regular intervals until the final message," said Cottam. "This was 'Come quick; our engine room is filling up to the boilers.'"

Cottam said that after the *Titanic*'s survivors were picked up he worked practically continuously until Tuesday, when he fell asleep at his post. He could not tell when he dropped from exhaustion nor when he awoke.

CHAPTER XXX

INVESTIGATION CONTINUED

BLUNDERS IN WIRELESS MESSAGES CONTRIBUTED TO GREAT LOSS OF LIFE

estimony given before the Senate committee showed that blunders in wireless service had much to do with the great loss of life.

Harold S. Bride, who was relief operator on the *Titanic*, said that when Chief Operator Phillips sent out the call for help the first answer came from the *Frankfurt* of the North German Lloyd line. The operator on the *Frankfurt* apparently considered the call trivial, for half an hour after receiving the imperative appeal he called the *Titanic* to inquire specifically just what was wrong.

"Mr. Phillips said he was a fool," Bride testified, "and told him to keep out, but did not tell him the *Titanic* was sinking."

No effort was made to re-establish communication with the *Frankfurt*, although Phillips felt certain the vessel was much nearer than the *Carpathia*, with which communication had been established.

◄ BRIDE IGNORED CALIFORNIAN'S CALL ►

Another phase of the laxity of the wireless service was developed when Chairman Smith drew from the witness an acknowledgment that Sunday evening Bride was sitting, the telephonic apparatus strapped to his ears, adjusting his accounts, while the steamship *Californian*, seeking to warn the *Titanic* that icebergs were invading the lanes of ocean travel, called incessantly.

Bride said he heard the call, but did not answer because he was "busy."

It was not until a half hour later that the *Californian*, striving to reach the steamship *Baltic*, reached also the *Titanic*, whereupon the warning that three great icebergs had been sighted was noted by Bride and verbally communicated to the *Titanic*'s captain.

⟨⟨ MARCONI CRITICIZED OPERATOR ⟩⟩

Senator Smith established by William Marconi that the *Titanic* and the *Frankfurt* operated virtually the same type of instruments.

Marconi also criticized the operator on the *Frankfurt* for neglecting to act immediately after he received the first call for help. It was the duty of the wireless operator, he said, to tell his captain of the distress signal so that that ship might have rushed to the rescue.

Both Bride and Thomas Cottam, wireless operators on the *Carpathia*, were mere boys, neither being over twenty-three years old.

Neither had any telegraph experience previous to taking up wireless telegraphy and both while on the stand told tales of long hours at low wages and days and nights spent without sleep.

This inexperience and the mental condition of the young operators were the two points on which Senator Smith bore persistently. He had put Cottam through a grueling examination, in which the youth testified that he had not slept more than eight or ten hours between Sunday night, when the *Titanic* called for help, and Thursday night, when the vessel docked. Bride's story bore out virtually all that Cottam's had established.

❦— TESTIMONY OF OPERATOR BRIDE —❦

"What practical experience have you had?" asked Senator Smith.

"I have crossed to the States three times and to Brazil twice," said Bride.

Bride remembered receiving and sending messages relative to the speed of the *Titanic* on its trial tests. After leaving Southampton on the *Titanic*'s fatal trip he could not remember receiving or sending any messages for Ismay. Senator Smith asked particularly about messages on Sunday.

"I don't remember, sir," said Bride. "There was so much business Sunday."

He was asked if Captain Smith received or sent any messages Sunday.

"No, sir," was the reply.

After testifying he made no permanent record of the iceberg warnings, Bride insisted he gave the memorandum of the warning to the officer on the watch. The name of the officer he could not tell. He did not inform Captain Smith.

❦— NOTIFIED OF THREE ICEBERGS —❦

Later the witness told of having intercepted a message from the *Californian* intended for the *Baltic*, which told of the presence of three giant icebergs in the vicinity of the former vessel.

"I gave the message to the captain personally," he said.

❦ MARCONI EXPLAINS "C.Q.D." ❦

In an effort to determine whether the signal "C.Q.D." might not have been misunderstood by passing ships Senator Smith called upon Mr. Marconi.

"The 'C.Q.,'" said Mr. Marconi, "is an international signal which meant that all stations should cease sending except the one using the call. The 'D.' was added to indicate danger. The call, however, now has been superseded by the universal 'S.O.S.'"

Senator Smith then resumed the direct examination of Bride, who has said the North German Lloyd was the first to answer the *Titanic*'s distress signal.

"Have you heard it said that the *Frankfurt* was the nearest to the *Titanic*?" the senator asked.

"Yes, sir; Mr. Phillips told me that."

"How did he know?"

"By the strength of the signals," said the witness, who added that the *Carpathia* answered shortly after.

In addition to further questions, Operator Bride said:

"We did not feel the shock when the ship struck. In fact, I was asleep and was not even awakened by the impact. When the engines stopped, Mr. Phillips called me and I put on the telephone apparatus while he went out to see what was the trouble. A little later he came back. He said things looked 'queer.' By 'queer' I suppose he meant that everything was not as it should be.

"When I heard the confusion on deck I went out to investigate, and when I returned I found Mr. Phillips sending out a 'C.Q.D.' call giving our position. We raised the *Frankfurt* first, and then the *Carpathia* and the *Baltic*. As I have said, we did not try for the *Frankfurt* for any length of time, but concentrated our messages on the *Carpathia*, which had answered that it was rushing to our aid.

"The captain came into the wireless cabin when the *Carpathia* advised us of its position and figured out the time when that vessel probably would arrive. He left when that was disposed of, and proceeded to the bridge. Then we began unofficially to keep in communication with the *Carpathia*.

"From time to time either Mr. Phillips or I would go on deck to observe the situation. The last time I went I found the passengers running around in confusion and there was almost a panic. They were seeking life belts. All of the large lifeboats were gone, but there was one life raft remaining. It had been lashed on the top of the quarters on the boat deck. A number of men were striving to launch it.

❦ Prepared for Ship's Sinking ❦

"I went back to the wireless cabin then. Mr. Phillips was striving to send out a final 'C.Q.D.' call. The power was so low that we could not tell exactly whether it was being carried or not, for we were in a closed cabin and we could not hear the crackle of the wireless at the mast. Phillips kept on sending, however, while I buckled on his life belt and put on my own. Then we both cared for a woman who had fainted and who had been brought into our cabin.

"Then, about ten minutes before the ship sank, Captain Smith gave word for every one to look to his own safety. I sprang to aid the men struggling to launch the life raft, and we had succeeded in getting it to the edge of the boat when a giant wave carried it away. I went with it and found myself underneath. Struggling through an eternity, I finally emerged and was swimming 150 feet from the *Titanic* when it went down. I felt no suction as the vessel plunged.

"Captain Smith stuck to the bridge, and, turning, I saw him jump just as the vessel glided into the depths. He had not donned a life belt, so far as I could see, and went down with the ship."

CHAPTER XXXI

THE INVESTIGATION IN WASHINGTON

HELP NEAR AT HAND, IGNORED DISTRESS CALL AND ISMAY'S ATTEMPTS TO GET BACK TO ENGLAND SHOWN

n the Senate investigating committee, April 22, Fourth Officer Boxhall made a startling revelation in regard to a ship close at hand at the time of the wreck which ignored all the *Titanic*'s signals. Also, in response to Senator Smith's questions he gave some evidence about the lifeboats. Boxhall said they had had a lifeboat drill before sailing in the presence of inspectors from the board of trade, in which only two boats on the same side of the ship were lowered. He declared that under the weather conditions at the time of the collision, the lifeboats were supposed to carry sixty-five persons. He said, too, that in accordance with the British board of trade regulations, the boats contained water breakers, water dippers, bread, bailers, masts, sails, lights and supplies of oil when the *Titanic* left Belfast, though he did not know if these things were in when the ship left Southampton.

◄ TESTIMONY OF OFFICER BOXHALL: LAUDED HABITS OF OFFICERS ►

Boxhall testified to the sobriety and good habits of his superior and brother officers.

"Lightoller was on the bridge when I came on at 8 o'clock. He was relieved at 10 o'clock by Mr. Murdock, who remained until the accident happened. Moody, the sixth officer, was on deck also. Fleland Leigh and the bridge officer, Mr. Murdock, were on the lookout," said Boxhall.

❦ Admitted Knowledge of Bergs ❦

Under questioning Boxhall said Captain Smith had told him of the position of certain icebergs which he had marked on the chart.

Senator Smith then asked the witness:

"Do you know whether the temperature of the water taken from the sea was tested?"

"Yes, sir; I saw the quartermaster doing it. He reported to the junior officer, Mr. Moody."

"Did you see the captain frequently Sunday night?" asked Senator Smith.

"Yes, sir; sometimes on the upper deck, sometimes in the chart room; sometimes on the bridge, and sometimes in the wheelhouse."

"Was the captain on the bridge or at any of the other places when you went on watch at 8 o'clock?"

"No, I first saw the captain about 9 o'clock."

❦ Ismay Not on Bridge ❦

"Did you see Mr. Ismay with the captain on the bridge or in the wheelhouse?"

"No, sir; not until after the accident."

Boxhall said he did not believe the captain had been away from the vicinity of the bridge at any time during the watch.

"When did you see the captain last?" asked Senator Smith.

"When he ordered me to go away in the boat."

"Where were you at the time of the collision?"

"Just approaching the bridge."

"Did you see what occurred?"

"No, I could not see."

"Did you hear?"

"Yes; the senior officer said 'We have struck an iceberg.'"

"Was there any ice on the deck?"

"Just a little on the lower deck. I heard the report of the crash."

"Did you see the iceberg?"

"No, sir."

First Officer Reported Accident

Boxhall then went to the bridge, where he found the first officer, Mr. Murdock; the sixth officer, Mr. Moody, and Captain Smith.

Boxhall said the captain asked what was the trouble, and the first officer replied they had struck an iceberg, and added that he had borne to starboard and reversed his engines full speed after ordering the closing of the water-tight doors.

"Did you see the iceberg then?"

"Yes, sir. I could see it dimly. It lay low in the water and was above as high as the lower rail of the ship, or about thirty feet out of the water."

Boxhall said he went down to the steerage, inspected all the decks in the vicinity of where the ship had struck, found no traces of any damage, and went directly to the bridge and so reported.

"The captain ordered me to send a carpenter to sound the ship," he said, "but I found a carpenter coming up with the announcement that the ship was taking water. In the mail room I found mail sacks floating about while the clerks were at work. I went to the bridge and reported, and the captain ordered the lifeboats to be made ready."

←⟨ Another Boat Nearby ⟩→

Boxhall testified that at Captain Smith's orders he took word of the ship's position to the wireless operators.

"What position was that?"

"41:46 north, 50.14 west."

"Was that the last position taken?"

"Yes, the *Titanic* stood not far from there when it sank."

After that Boxhall went back to the lifeboats, where there were many men and women. He said they had life belts.

"After that I was on the bridge most of the time, sending out distress signals, trying to attract the attention of boats ahead," he said. "I sent up distress rockets until I left the ship, to try to attract the attention of a ship directly ahead. I had seen its lights. It seemed to be meeting us, and was not far away. It got close enough, it seemed to me, to read our electric Morse signals. I told the captain. He stood with me much of the time trying to signal this vessel. He told me to tell it in Morse rocket signals, 'Come at once—we are sinking.'"

←⟨ Saw No Answering Signal ⟩→

"Did any answer come?" asked the senator.

"I did not see them, but two men say they saw signals from that ship."

"How far away do you think that ship was?"

"Approximately five miles."

Boxhall said he did not know what ship it was.

"What did you see on the ship?"

"First we saw its masthead lights, and a few minutes later its red side lights. It was standing closer."

"Suppose you had had a powerful searchlight on the

Titanic, could you not have thrown a beam on the vessel and have compelled its attention?"

"We might."

❦ Rowed About After Wreck ❦

Boxhall said he had rowed in the seaboat three-quarters of a mile when the *Titanic* went down. Before that he had rowed around the ship's stern to see if he could not take off three more persons for whom there was room. He abandoned that attempt, however, because he had with him only one man who knew how to handle an oar and he feared an accident. His boat, he said, was the first picked up by the *Carpathia*. That was about 4:10 in the morning.

"Did you have any conversation with Mr. Ismay that night?"

"Yes, sir, before I left the ship. On the bridge just before the captain ordered me below to take an emergency boat."

"When you boarded the *Carpathia*, did you see any lights on any other lifeboats?"

"No. It was nearly daylight. It was daylight by the time I got my passengers aboard the *Carpathia*."

"Could you say any other lifeboats had lights besides yours?"

"I saw several with lanterns. These lanterns were beside the helmsman in each case and on the bottom of the lifeboats. I would not say all the boats had lights."

❦ Saw Ismay in Lifeboat ❦

Boxhall said he knew none of the American passengers personally, but he knew the identity of Colonel John Jacob Astor.

"Did you see Ismay when you got into the lifeboat?"

"No."

"When did you next see Ismay after you left the ship?"

"I saw him in a collapsible boat afterward."

"Any women in it?"

"Yes, it was full of them—well, not exactly full, but there were many women—most of them foreigners."

"How long after you reached the *Carpathia* did Ismay's boat arrive?"

"I cannot say exactly, but it was before daylight."

❊⊰ Saw None Refused Rescue ⊱❊

Boxhall heard persons on the *Titanic* say some people refused to enter the lifeboats, but he saw no one ejected from the boats, nor prevented from entering.

"Did you see any who got in from the water or see any in the water?"

"No, sir," said Boxhall. "If I had seen any in the water I should have taken them in the boat."

Boxhall said the sea was calm and that in his opinion each of the lifeboats could have taken its full capacity. How many he had in his small seaboat he never knew.

Senator Newlands returned to the subject of the icebergs.

"You say you could not see these giant icebergs when in the seaboat, but you could hear the water lapping against them?"

"Yes, sir. It was an oily calm and we could see nothing in the small boats."

"If the sea is smooth, then, it is difficult to discern these icebergs?"

"Yes, sir. I believe if there had been a little ripple on the water the *Titanic* would have seen it in time to avoid it."

⫷⫸ Testimony of Franklin ⫷⫸

P.A.S. Franklin was the next witness called. Mr. Franklin described the business operations and extent of the International Mercantile Marine.

"What is its capitalization?" asked Senator Smith.

"One hundred million in common and preferred shares, $52,000,000 in 4½ percent bonds, $19,000,000 in 5 percent bonds and about $7,000,000 of underlying bonds."

After Mr. Franklin had read a list of the officials and directors of the International, Senator Smith said:

"Did you know Captain Smith of the *Titanic*?"

"Ever since 1898," said the witness, adding that Capt. Smith had commanded the *Majestic*, *Adriatic*, *Baltic*, *Olympic* and the *Titanic*.

⫷⫸ No Message from Smith ⫷⫸

"So far as you know, did you or any of your subordinate officers have any communication with Captain Smith on his last voyage?"

"None at all."

Mr. Franklin said he had received no communication from Mr. Ismay except one by cable from Southampton. This, he said, was merely a cablegram announcing the complete success of the *Titanic*'s trial trip and favorable prospect for a successful voyage.

Senator Smith then showed Mr. Franklin the telegram received by Congressman Hughes of W. Va. from the White Star line, dated New York, April 15, and addressed to J. A. Hughes, Huntington, W. Va., as follows:

> "*Titanic* proceeding to Halifax. Passengers probably land on Wednesday. All safe.
>
> The White Star Line"

"I ask you," continued the senator, "whether you know about the sending of that telegram, by whom it was authorized and from whom it was sent?"

"I do not, sir," said Franklin. "Since it was mentioned on Saturday we have had the entire passenger staff examined and we cannot find out."

❖ First Warning of Tragedy ❖

Asked when he first knew the *Titanic* had sunk, Franklin said he first knew it at 6:27 P.M. Monday. He then produced a thick package of telegrams which he had received Monday in relation to the disaster.

"How did you ascertain the location of the *Olympic*, *Baltic*, and others?" asked the senator.

"We worked them out on our charts. We had no direct communication from any of the ships. Our first endeavor to communicate with our big ships was a message sent April 15 at 3 o'clock A.M. This message read as follows:

"'Haddock, *Olympic*: Make every endeavor to communicate *Titanic* and advise position and time. Reply within the hour.'"

❖ Message Sent to *Olympic* ❖

Franklin said the *Olympic* was dispatched this message:

"Haddock, *Olympic*: Rumored here *Titanic* sunk. Cannot confirm here. Expect *Virginia* alongside. Franklin."

"At 6:20 or 6:30 Monday evening," Mr. Franklin continued, "a message was received telling the fateful news that the *Carpathia* reached the *Titanic* and found nothing but boats and wreckage; that the *Titanic* had foundered at 2:20 A.M. in 41.16 north, 50.14 west; that the *Carpathia* picked up all the boats and had on board about 675 of the *Titanic*'s survivors, passengers and crew. This message was from Haddock also.

"After that we got another message from Haddock stating that 'Yamsi,' meaning Ismay, was on the *Carpathia*."

❦ Messages Ended Hope ❦

One by one Mr. Franklin read telegrams that had been hurled through the air from shore to the ships and from them back to the shore. All hope that some other vessels besides the *Carpathia* had picked up some of the *Titanic*'s survivors was dissipated when the *Olympic* flashed word that neither the *Baltic* nor the *Tunisian* had any of the *Titanic*'s people aboard.

Senator Smith sought to discover who had been "tampering with the wireless operators or had been responsible for the failure of the wireless to get the news to shore earlier." Mr. Smith repeatedly asked the witness whether he had not had a conference Monday morning with Mr. Marconi or Mr. Sammis, chief engineer for the Marconi company.

"No, most emphatically," said the witness. "In no way did I attempt or cause to be attempted any censorship of the wireless."

"Did you receive at any time from any one or any officer of your company a request that the steamship *Cedric* be held at New York until the arrival of the *Carpathia*?" Senator Smith asked.

"Yes, sir," said the witness, and began to read a telegram from the *Carpathia*.

"What time was it received?"

"At 5:19," said the witness, who said the telegram asked that the *Cedric* be held because the sender considered it "most desirable" that the members of the crew be sent back on the

Cedric and declaring his intention of sailing on that ship himself. The sender also asked that clothing and shoes be brought to the dock for him when the *Carpathia* got in.

❧ Ismay Signed Cipher ❧

"By whom was that signed?" asked Senator Smith.

"Yamsi."

"Do you know who Yamsi is?"

"Yes, sir. It is cipher for Mr. Ismay's signature. I sent in reply the following:

"'Yamsi, *Carpathia*: Have arranged forward crew *Lapland*, sailing Saturday, calling at Plymouth. We all consider most unwise to delay *Cedric* considering circumstances.'"

Senator Smith then had Franklin read all the messages that passed between himself and Ismay on the *Carpathia* April 18. At 5:30 A.M. of that day Franklin received from Ismay this message: "Send responsible White Star ship officer and fourteen men to two boats to take charge of thirteen *Titanic* lifeboats at quarantine."

Franklin testified that he received a message from Ismay on the *Carpathia* a little later on the morning of the 18th to join the *Carpathia* at quarantine and that several other messages came from him urging that the *Cedric* be held. After all these had come in Franklin cabled Ismay:

"Think it most unwise to retain *Cedric* in New York." This was followed by a reply from Ismay which included: "Unless you have good and sufficient reason to hold the *Cedric*, kindly do so."

❧ Learned of Senate Inquiry ❧

In an effort to connect the attempted departure of Mr. Ismay and the *Titanic* crew with the Senate's investigation, Senator Smith asked the witness when he had learned the Senate had decided to investigate the disaster.

"I think about 2 o'clock Thursday."

"Did you communicate the information to your company?"

"I did, that night, by cable, I think."

"When did you advise Mr. Ismay?"

"I told him of it when I got aboard the *Carpathia*," said the witness.

Senator Perkins took Mr. Franklin in hand and questioned him at some length as to the safety equipment of the *Titanic*.

The *Titanic's* equipment was in excess of the law," said the witness. "It carried its clearance in the shape of a certificate from the British board of trade."

❦ Safeguards on Other Ships ❦

Senator Bourne took up the same line of questioning.

"Has anything been done with the equipment of other ships as a result of the disaster?" he asked.

"Most emphatically," answered Mr. Franklin. "On last Friday Mr. Ismay ordered that all our vessels be equipped with boats and rafts sufficient to take off every passenger and every member of the crew in case of accident."

"Do you know of any one, any officer or man, or any official who you deem could be held responsible for the accident and its attendant loss of life?"

"Positively not. No one thought such an accident could happen. It was undreamed of."

Mr. Franklin volunteered a statement relating to criticisms of the White Star Company for attempting to return the crew of the *Titanic* to Europe immediately.

"I think there has been an awful mistake made about that matter," said Franklin. "I would like to clear it up. The criticisms have been made that we were trying to keep those men from testifying. That is not so. It was not the reason at all. As far as the crew are concerned it was our duty to return them to their homes. We assured you that we would hold any officers or men that you wanted for this committee."

Senator Newlands brought out that the speed of the *Titanic* at the time of the accident was about four miles an hour below that of the *Mauretania* and *Lusitania*.

"Do you have rules governing the running of a ship in fog or when ice is in a ship's vicinity?"

"We have stringent rules. None of the commanders that I have ever had communication with ever got the idea from me that our company wanted records broken."

CHAPTER XXXII

SENATE COMMITTEE EXAMINED LOOKOUT AND PASSENGERS

FURTHER DESCRIPTION OF THE WRECK BY AN EYEWITNESS IN OFFICIAL TESTIMONY — MARINE GLASSES FOR LOOKOUT MIGHT HAVE PREVENTED WRECK

ailure to provide binoculars or spy glasses for the lookouts on the *Titanic* was one contributing cause of that ship's loss and, with it, the loss of 1,600 lives. Two witnesses before the Senate investigating committee agreed on this. They were Frederick Fleet, a lookout on the liner, and Maj. Arthur Godfrey Peuchen, Canadian manufacturer and yachtsman, who was among the rescued passengers.

◄― MIGHT HAVE AVOIDED BERG ―►

Fleet acknowledged that if he had been aided in his observations by a good glass he probably could have spied the berg into which the ship crashed in time to have warned the bridge to avoid it. Major Peuchen also testified to the much greater sweep of vision afforded by binoculars and, as a yachtsman, said he believed the presence of the iceberg might have been detected in time to escape the collision had the lookout men been so equipped.

It was made to appear that the blame for being without glasses did not rest with the lookout men. Fleet said they had asked for them at Southampton and were told there were none for them. One glass, in a pinch, would have served in the crow's nest.

⪡ LACKED EXPERIENCED SAILORS ⪢

Major Peuchen criticized in strong terms the lack of experienced sailors on board the *Titanic*. He said that when the call to quarters was sounded not enough of the crew responded to undertake the work required in lowering and filling the boats. Furthermore, he said, no drills had been held from the time the ship left Southampton, although it was customary to hold such drills every Sunday.

Herbert J. Pitman, third officer of the *Titanic*, told of his failure to turn back the lifeboat in which he and his passengers were idly drifting, to attempt the rescue of others when the *Titanic* went down. Shuddering at the recollection, he said the cries for help made "one long, continuous moan."

The passengers insisted that to go back to aid would mean their destruction, he said, so that after starting in the direction of the cries he rescinded his orders and waited for the dawn. Twice he begged to be spared a recital of the facts, but Senator Smith pressed him.

⪡ ISMAY KEPT IN CAPITAL ⪢

J. Bruce Ismay, managing director of the International Mercantile Marine, and Vice President P. A. S. Franklin of the White Star Line, urgently requested the committee to permit them to return to New York.

In executive session at the close of the hearing the committee declined to allow either to leave Washington until he was no longer needed.

❦ PHOTOGRAPHERS DRIVEN OUT ❦

The importunities and activities of a squad of photographers so aroused Senator Smith that he indignantly ordered them all excluded from the chamber.

"This inquiry is official and solemn," he said in explanation, "and there will be no hippodroming or commercializing of it. I will not permit it."

An amateur photographer managed to slip past the guard later, but was summarily ejected when he sought to get a snap of the scene.

❦ CROWD EXCLUDED ❦

Owing to the constant interruptions during the interrogation of witnesses the Senate committee determined to exclude the general public. To accomplish this the hearing was transferred to a smaller room in the Senate office building. Only witnesses, those particularly interested in the inquiry and members of the press were admitted to the room.

Herbert J. Pitman, third officer, was the first witness of the day. It had been expected that J. B. Boxhall, fourth officer, would be recalled, but it was announced he was ill.

❦ ONLY SIXTEEN MEN DRILLED ❦

Pitman said that in the boat drill conducted by the board of trade at Southampton approximately eight men went in each of the two boats used in the drill.

The witness maintained that virtually the only way to discover the proximity of icebergs was to see them, asserting that, while science may hold there are numerous ways, they never have been demonstrated.

Pitman was on the bridge of the *Titanic* from 6 to 8 o'clock the night of the collision. After that he went to his berth. Half asleep at the time of the accident, he said he wondered sleepily where they were anchoring. It was nearly time for the next watch, so he dressed leisurely and was lighting his pipe when Mr. Boxhall told him the ship had struck an iceberg. He went forward and saw ice, and then walked back, where a number of firemen coming up told him there was water in the hatch.

◀◀ ISMAY REALIZED PERIL ▶▶

Going on deck he met a man whom he afterward learned was Mr. Ismay, who said, "Hurry, there's no time for fooling." Mr. Ismay helped him load the boat in which Pitman embarked on orders from Mr. Murdock after calling for more women passengers and finding there were none in sight.

The witness said that just before the boat pulled away Mr. Murdock leaned over, shook his hand, and said, "Goodbye and good luck, old man."

"When you shook hands with Murdock did you expect to see him again?"

"Certainly."

"Do you think he expected to see you again?"

"Apparently not, but I expected fully to be back on the ship in a few hours."

◀◀ GOING AT FULL SPEED
WHEN BERG WAS STRUCK ▶▶

Pitman told of the placing on the chart of crosses indicating the presence of icebergs by the fourth officer and said that the speed had been increased from twenty and one-half knots on leaving Southampton to twenty-one and one-half knots and that he supposed the ship was going at top speed when it struck.

The witness said he had not seen any Morse signals on the *Titanic* and did not of his personal knowledge know of the presence of another ship, but that he later had heard that one had passed.

❈⊷ SOUNDED WARNING OF BERG ⊶❈

Fleet said that he went into the crow's nest at 10 o'clock and, obeying a warning, kept a sharp lookout for ice. At 11:30 o'clock he reported a black mass ahead, but could not tell how long it was before the collision came. He sounded three bells and telephoned to the bridge that there was an iceberg ahead, and soon the ship started to turn to port.

When Fleet first saw the berg it appeared about the size of two big tables, he said, but when struck it proved to be fifty or sixty feet high.

Fleet said that when the collision came there was little impact and "just a sharp grinding noise."

"Did it alarm you?" asked the senator.

"No, I thought it was a narrow shave."

❈⊷ HAD NO SPY GLASS ⊶❈

"Did you have glasses?" asked Senator Smith.

"No, sir."

"Isn't it customary for the lookouts to use glasses in their work?"

"Yes, sir, but they didn't give us any on the *Titanic*. We asked for them at Southampton, but they said there were none for us."

❈⊷ COULD HAVE ESCAPED ⊶❈

"We had a pair from Belfast to Southampton, but none from

Southampton to the place of the accident."

"What became of the glasses you had from Belfast?"

"We do not know."

"If you had had glasses could you have seen the iceberg sooner?" asked Senator Smith.

"We could have seen it a bit sooner," said Fleet.

"How much sooner?"

"Enough to get out of the way."

"Were you and Leigh disappointed that you had no glasses?"

"Yes."

"Did the officers on the bridge have glasses?"

"Yes."

❈⤙ MAJOR THOUGHT SHOCK WAVE ⤚❈

Major Peuchen was the first passenger witness to appear before the committee. All ten of his friends with whom he was traveling lost their lives in the wreck. The major told of the trip and said:

> "There was no mention of fire and we were all pleased with the trip until the crash. After 11 o'clock I went to my stateroom. I scarcely was undressed when I felt a shock. I thought merely that a large wave had struck the ship. Fifteen minutes later I met Charles M. Hays of the Grand Trunk-Pacific. I asked him, 'Have you seen the ice?' He said 'No.' Then I took him up and showed him. Then I noticed the boat was listing. I said to Mr. Hays:
>
> "'It's listing; it shouldn't do that.'
>
> "He said: 'Oh, I don't know. This boat can't sink.' He had a good deal of confidence and said: 'No matter what we have struck it's good for eight or ten hours.'

❦ Seemed Short for Sailors ❦

"I met my friend Beattie, who said: 'The order is for the lifeboats. It is serious.' I couldn't believe it at first, but went to my cabin and changed to some heavy clothes."

The witness said when he got on deck the boats were being prepared for lowering on the port side.

"They seemed to be short of sailors around the lifeboats where I was. When I came on deck first it seemed to me that about one hundred stokers came up with their gunny sacks and crowded the deck. One of the officers, a splendid man, drove these men right off the deck. He drove them like sheep."

❦ Called Smith Attentive ❦

"Do you think the captain was attentive to his duties?"

"Yes, I do."

Major Peuchen praised the women who rowed in the boats. He said there was room in some boats which left from the port side and he could not understand why more men were not taken off.

❦ No General Alarm Sounded ❦

Several senators asked if the fact that there was no general alarm sounded after the collision might account for the failure of many women to appear on the decks in time for the lifeboats. He thought that was probable.

Major Peuchen told the committee he thought that if the lookouts on the *Titanic* had glasses the ship might have been saved from the collision.

CHAPTER XXXIII

MEMBERS OF SHIP'S CREW ON STAND

MEMBERS OF THE CREW TOLD THEIR STORIES OFFICIALLY, DESCRIBING FOR THE MOST PART THE LOADING OF THE LIFEBOATS AND THE CONDUCT OF ISMAY

arold G. Lowe, fifth officer of the *Titanic*, told his story of the wreck before the investigation committee. His testimony revealed the fact that, with a volunteer crew, he rescued four men from the water, saved a sinking collapsible lifeboat by towing it and took off twenty men and one woman from the bottom of an overturned boat, all of whom he landed safely on the *Carpathia*. Lowe testified that he looked over the lifeboats in Belfast Harbor and found everything in them, except a dipper which was missing from one. He was not sure whether a fire drill had been held or not. He did not know whether the officers were at their right places on the side of the ship where he was or not. He was not on duty Sunday night and could not be induced to make a positive statement of the ship's position, though he had a memorandum of the speed on that day as a fraction below 21 knots an hour. He asserted that he was a temperate man.

◈◄ TESTIMONY OF OFFICER LOWE ►◈

The witness said he did not know when he was awakened. He said he dressed hurriedly and went on deck and found people with life belts on the boats being prepared. He began working at the lifeboats.

"I was working the boats under First Officer Murdock," he said. "Boat No. 5 was the first one lowered.

"There were about ten officers helping, two at each end, two in the boat, and others at the ropes."

⊹ Ordered Ismay to Keep Quiet ⊹

"A steward met me on the *Carpathia*. He said to me, 'What did you say to Ismay that night on the deck?' I said that I did not know that I had said anything to Mr. Ismay. I did not know him. Well, the steward on the *Carpathia* said I had used strong language to Mr. Ismay. I happened to talk to Ismay because he appeared to be getting excited. He was saying excitedly, 'Lower away, lower away, lower away.'"

Chairman Smith asked Mr. Ismay about the language and Mr. Ismay suggested that the objectionable language be written down to see if it was appropriate. This was done. They returned to the question of lifeboats after Lowe explained that Ismay "was interfering with our work. He was interfering with me, and I wanted him to get back so that we could work. He was trying to get in the boat."

⊹ Denied Talking with Women ⊹

"How many men were in the boat?"

"I'm not sure, sir, but I should say about ten."

Lowe denied having conversed with Mrs. Douglas or Mrs. Ryerson on board the *Carpathia*.

Senator Smith asked Lowe if in his opinion the lifeboat before it was lowered was loaded to its proper capacity.

Lowe tried to avoid making a direct answer. Senator Smith insisted upon an answer.

"Yes, sir," said Lowe, finally, "I think it was properly loaded for lowering."

"What is the official quota for such a lifeboat?"

"It can carry sixty-five adults and say, a boy or girl."

"Then you wish the committee to understand that a lifeboat under British regulations could not be lowered with safety with new tackle and equipment containing more than fifty people?"

"The dangers are if you overcrowd the boat it will buckle up from the two ends," said Lowe. "The 65.5 is a floating capacity. If you load from the deck to lower I should not like to put more than fifty in a lifeboat."

Senator Smith referred to Third Officer Pitman's testimony in which he said there were thirty-five persons in lifeboat No. 5. That being the case, he asked why Pitman could not have gone to the rescue of the drowning, whose cries he heard plainly, but did not heed.

"Had he attempted to rescue those in the water he would have endangered the lives of those with him," Lowe asserted.

❦ Denied Lack of Oarsmen ❦

Senator Smith asked if it were not true that the reason why the boats were not properly loaded was because the crew were not able to row. The witness denied this.

"What was the drill for at Southampton?" asked the chairman.

"It was for the board of trade."

"There were eight men to a boat then. They were all oarsman. Where were they when you were loading lifeboat No. 5?

"You must remember, sir, we were in harbor and we had

the pick of the men. At the time of the collision the men went down with the 'bosun' to clear away the gangway doors to make way for the loading."

The witness said the discipline was excellent. Only one boat, a collapsible one, overturned.

Senator Smith asked the number of the crew and the witness said so far as he knew there were 903 of them.

"And with 903 men aboard," said the senator, "you did not have enough to man twenty lifeboats properly?"

The witness demurred and the chairman showed his disapproval, going to the extent of criticizing the officer's refusal to make direct replies.

❮ Did Not Refuse Anyone ❯

Senator Smith then sought to discover whether any men, women, or children had been refused admission to the boats or were put out of the boats after they had gotten in. The officer said no one was refused and declared the only confusion was by the passengers interfering with the lowering gear.

"There was no such thing as selecting. First we took the women and children, then others as they came. There was a procession at both ends of the boat; in little knots they were, little crowds."

"Was Mr. Ismay there?"

"Yes, he was; he was right alongside of me. I didn't know it was Mr. Ismay then, but I know now. It was the same man whom I had ordered not to interfere in lowering No. 5. But he took hold and was helping afterward. I could see his face in the glare of the rockets, and he aided in lowering boat No. 3."

Lowe told of tying five of the lifeboats together, transferring the passengers from his boat, and then called for volunteers to row back to the wreck.

"We rowed back and around the wreck," said the witness, "and we picked up four men who were struggling in the water."

"You said a moment ago that you had waited before returning to the wreck until 'things quieted down,'" said Senator Smith. "What did you mean by 'quieted down?'"

"Until the cries ceased."

"The cries of the drowning?"

"Yes, sir. We did not dare go into the struggling mass. It would have sunk us. We remained on the edge of the scene, but it would have been suicide to have gone in."

"How long did it require for things to get quiet?"

"About an hour and a half."

"How many persons were on your boat when you went alongside the *Carpathia*?"

"About forty-five. I took them off a sinking collapsible boat. I left the bodies of three men."

Senator Smith wanted to know about the shooting on board the *Titanic* while it was sinking. Lowe said he had fired three shots into the water to scare away some immigrants on one of the decks who he feared were about to swamp a loaded boat by jumping. He was certain the shots struck no one.

◀— TESTIMONY GIVEN BY LIGHTOLLER —▶

Chief interest in the testimony of C. H. Lightoller, second officer of the *Titanic*, was centered in his story of the actions of J. Bruce Ismay.

Senator Burton asked the witness to relate his conversation with Ismay on the *Carpathia*. Lightoller said he and his brother officers talked over the sailing of the *Cedric* and had agreed it would have been a "jolly good idea" if they could catch the vessel. It would result in keeping the men together and let everyone get home.

The spirit of heroism.

"Mr. Ismay, when the weather thickened, remarked to me," said Lightoller, "that it was hardly possible that we could catch the boat. He asked me if I thought it desirable that he send a wireless to hold the *Cedric*. We were all agreed that it was the best course and we all advised it."

❦ Ismay Deplored Rescue ❧

"I would say that at that time Mr. Ismay was in no mental condition to transact business," said Lightoller. "He seemed to be possessed with the idea that he ought to have gone down with the ship because there were women who went down. I tried my best to get that idea out of his mind, but could not. I told him that there was more for him to do on earth and that he should not let the idea possess him that he had done a wrong in not staying back to drown. The doctor on the *Carpathia* had trouble with Mr. Ismay on the same ground.

"I was told on the *Carpathia* that Chief Officer Wild, who was working at the forward collapsible boat, told Mr. Ismay there were no more women to go. Ismay still stood back and Wild, who was a big, powerful man, bundled him into the collapsible boat."

Senator Smith asked Lightoller why when he testified in New York he did not tell about the sending of the telegram from the *Carpathia* urging that the *Cedric* be held.

"I did not say anything about it then because there had been nothing said about the telegram at that time," said Lightoller.

"Did you know when you sent the message the Senate was going to hold an investigation?"

"Most certainly not, or the telegram would never have been sent. Our sole idea was to keep witnesses together for just such an investigation, which we knew would be made in England."

Lightoller said that S. Hemmings, a lampman, who was waiting to testify before the committee, walked the length of the ship just before it sank and had seen only two women.

CHAPTER XXXIV

THE BEREFT IN THE BOATS

BY FRED S. MILLER

n the first stories of the *Titanic* disaster broadcast by the press of two continents, the obvious and spectacular features were of course most emphasized. Sensational columns-full lauded the heroism of the hundreds dead, and told the chiefest incidents of the wreck; then came shrieking denunciation of the shipowners, as their recklessness was revealed in the senatorial inquiry. And now that all the facts are known, the account bids fair to stand thus in men's minds: for the heroes, praise to the skies; damnation for the guiltily responsible, whose laxity or greed brought about the tragedy.

One item is too little dwelt upon. Although we judge unsparingly all criminal carelessness, and while we fittingly remember those who gave their lives to rescue others, we owe a tender duty also to the rescued, who were hurried over the vessel's side amid the midnight agony and uproar—goodbyes said in the sudden bewilderment of terror about which rang the fearful summons "Women and children first!"

At this it had taken much manly authority to induce these wives to be saved, also (glory of humanity!) a deal of lying.

"It's best for you to get in the boat, dear, though of course there's no real danger in my staying here! The *Titanic*'s unsinkable, you remember. Captain Smith wants all the women and children—why just think of ours!—away, so as to be on the safe side, that's all. There's another steamer coming, and when it picks you up in the morning you'll find me right here!"—

And so forth. Those husbands, how they laid it on. "Men were deceivers ever!" Thus they stayed a panic; doing all that inexperienced brave men could do in that crisis of the wreck to turn a few scant boatsful from the yawning gulf into which the ship was every instant sinking, sinking.

So the women and their little ones were hurried to the rail and lowered to the blackness far beneath. Rowing away, they could turn their eyes to the steamer which yet showed no evidence of collapse, as it loomed across the water, its huge hulk outlined quite from end to end by rows of glowing lights—when on an instant these lights faded sicklily, then died! As though to shut from those who longingly looked back a last faint ray of hope, left as they were now quite cut off, adrift in the unutterable profound. Beneath, two-thousand fathoms-deep of heaving ocean, over which they poised buoyed only the boat's inch planks; above, the deeper depth, black midnight far as the illimitable stars.

All sense of distance and direction speedily was lost for them; we may imagine the awed conjectures:

"Where is the vessel?"

"Over there, very dimly seen—so far we must have come!"

"But what is that other shape? How strange, a huge hill rising awful in the sea!"

"No! the iceberg on which the steamer struck."

"I had thought the *Titanic* would have shivered anything of ice; yet there the berg uprears itself unmoved, as though it lingered patient to see the end!"

Also we may imagine that they comforted one another and soothed the wailing children, as is the wont of women; prayed to the good God and were heartened so—prayers for the safety of the stricken ship yet faithful to its trust of keeping safe their loved ones.

So they drifted, an hour in the chill northern night, suffering intensely, seeing nothing but their own dim huddled forms, hearing nothing but a faint, confused, deceiving murmur from the vessel, and the harsh grinding of the ice cakes littering the ocean all about. It had been the captain's orders that they keep to the boats; they would do their duty—never mind the cold—blindly obey—theirs not to reason why! Joy cometh in the morning; and when the blessed light should prove the fear of wreck had only been a temporary vague alarm, they would row back to where—each felt assured—was one who longed for her as she was longing now. Saved from the sea, then; reunited! Never to be parted more!

Who may conceive their feelings when with a horror of amazement the explosion came, and sheets of fire sent soaring from the steamer's funnels revealed to land and sea that all was lost. When the pierced monster, with a rending roar, reared its prodigious bulk full upright in the ocean, poised so for an instant and then plunged, quenching all hope, leaving the waste of waters blacker with despair and night. We may believe that none of the terror of the scene was felt by those to whom it brought an overwhelming desolation. They were not appalled—no more than were those other women when "there was darkness over all the land until the ninth hour," when the rocks were rent and the graves opened. Perhaps they were awed by the contemplation of a sacrifice, for the first time comprehending why the men gave up their places in the boats; perhaps they were stricken numb with a grief too great for tears.

And would that that were all! For thence the night brought forth a crueler infliction. What had been, was frightful; but what ensued was an exquisite torture for the pitiful unoffenders, forced to hear the agony of those drowning, who moaned amid the lacerating ice cakes, cried with a loud voice and yielded up the ghost, or called again beseeching help where help was none. Help? "The depth saith it is not in me, and the sea saith it is not with me!" Those in charge of the boats returned as pitiless a silence. Although the women begged, they dare not venture back among the gasping hundreds battling desperately with death amid the icy waves.

For an hour the dying cries kept on—a long, intolerable and agonizing hour, a blended hum of multifarious woe upwelling from the waters, a mystery of awful utterance in the blackness of the night. How it smote on those who could not save! Also there were other voices, right at hand, as here:

"Oh, Mamma, listen! That's Papa! I hear him calling, calling! Why don't the men row back? It's so cold for him in the water!"

"We can't go back after those stiffs!" is the answer of a boatswain, as sworn to in the Senate's inquiry. A man can be more callous than the elements; not even the iceberg's adamant can match that piece of netherstone, his heart.

How wives and mothers listened yet endured it all may never be described. Mercifully only one went mad. Also by mercy's grace the rest, with gratitude unbelievable, could note the mounting quiet as the moans grew less and the deep claimed its sacrifice of saviors. Finally all were gone—not a gasp, not another choking sigh—the offering accepted, the immolation made complete; with the sea laid smooth again and swept with the proclaiming breeze and the minutest first faint light streaks of the dawn.

Then o'er the waves came humankind bringing rescue, bringing the love and outpoured pity of the world of men.

Doubtless, human sympathy is the divine consolation. That they could bring the story of that midnight to the universal heart, laying thereon the sacrifice of their heroic dead—this privilege soothed away, for the bereft in the boats and for their pitying friends world 'round, the dark and blighting aspects of a tragedy unhuman and terrific. For we are all fellow partakers of a reverence for unselfishness; we all hunger and thirst after the righteousness of saviors; and we are all allied against unpitying nature, sharing the yoke of domineering chance and change—bound in affection so.

Thus is preserved, from all the wreck of the *Titanic*, only the memory of an exalted offering. Quickened, also, the assurance that man is, somehow, kin to the Giver of every good and perfect gift.

This assurance persists, triumphant over man's every overthrow by his adverse environment. Whence comes it, in despite of the despairing, harsh vicissitudes that torture and perplex their puppet here, affirming at each unmerited assault—there is no God! It springs from human kindness; it is born of our mutual helplessness and our reliance on each other; confirmed by deeds of devotion and the reverence that accepts them. By the hour-long sacrificing death in icy waters, by the anguish of the ones who hovered near but were too weak to save.

So is revealed humanity's refuge and strength, called by them of old time "the fear of the Lord." Our privilege is to recognize it in every helpful act, in every kindly thought. Yea, in manifold nature also it is our highest wisdom to perceive it, even when her mysterious climaxes seem to laugh all human effort, faith and trust to scorn; when the pitiless depth saith it is not in me, and the angry sea saith it is not with me!

CHAPTER XXXV

TITANIC'S DEAD BROUGHT BACK

RETURN OF THE FUNERAL SHIP MACKAY-BENNETT WITH THE BODIES OF THE 190 VICTIMS OF THE DISASTER PICKED UP AT SEA

 y arrangement with the officers of the White Star Line, the cable ship *Mackay-Bennett* was dispatched to the scene of the disaster to pick up as many of the bodies of the victims as possible. She returned to Halifax, N.S., on April 30, leaving another vessel, the *Minia*, to continue the gruesome search.

Steaming slowly into Halifax harbor the *Mackay-Bennett* reached her dock in the navy yard shortly after 9:30 A.M., while the city's church bells tolled and British flags fluttered at half mast.

It was announced that the total number of bodies on board was 190 and it had been found necessary to bury 116 at sea. Among those brought back to port were the bodies of two women.

✦ ASTOR'S BODY TAKEN TO MORGUE ✦

Colonel Astor's body was taken from the ship shortly before noon and borne with others to the morgue.

Capt. F. H. Larnder described the work of the *Mackay-Bennett* at sea. The number of bodies found, he said, was 306. Of these 116, most of them members of the *Titanic*'s crew and unidentified, were consigned to the sea. Only 18 bodies of women were found afloat.

Relics of the great *Titanic* dotted the sea over an area thirty miles square, Captain Larnder said. Doors, windows and chairs by the score were found floating, but to none of them were bodies lashed. In several instances there were groups of floating bodies numbering fifty or more. Colonel Astor was found almost erect in his lifebelt.

Small boats were lowered by the *Mackay-Bennett* whenever a group of bodies was sighted, and into these the dead were piled three or four at a time. Hauled on board the cable ship, each was numbered with a large canvas tag and the valuables and papers were placed in a canvas sack similarly numbered.

❦ CONDUCT SERVICES FOR DEAD ❦

Canon K. O. Hind of All Saints' Cathedral, Halifax, who was on board, conducted the services in connection with the burial at sea. On three separate occasions services were held.

"We buried so many at sea," said Captain Larnder, "simply because we could not accommodate them. We had limited embalming supplies, and it was necessary to consign many to the deep. The majority of those sunk were unidentified. We had instructions when we left here to pick up all the *Titanic*'s dead we found, but under the conditions it was impossible to carry out these instructions."

It was announced that there was no doubt of the identification of Colonel Astor's body. In the pockets $2,500 cash had been found and he wore a belt with a gold buckle. The body identified as that of Mr. Widener was buried at sea.

The bodies were all tenderly and respectfully cared for. Those identified were delivered to relatives or friends and the unidentified were given Christian burial at Halifax, whose citizens purpose erecting a monument to their honored memory.

Relatives of *Titanic* disaster victims.

❦⊰ TWO GREAT NATIONS MOURN ⊱❧

When the news of the disaster to so many noted British subjects and American citizens was received, messages of condolences were exchanged by King George of England and President Taft as follows:

❦⊰ King George's Message ⊱❧

"The Queen and I are anxious to assure you and the American nation of the great sorrow which we experienced at the terrible loss of life that has occurred among the American citizens, as well as among my own subjects, by the foundering of the *Titanic*. Our two countries are so intimately allied by ties of friendship and brotherhood that any misfortunes which affect the one must necessarily affect the other, and on the present terrible occasion they are both equally sufferers.

— George R. and I"

❦⊰ President Taft's Reply ⊱❧

"In the presence of the appalling disaster to the *Titanic* the people of the two countries are brought into community of grief through their common bereavement. The American people share in the sorrow of their kinsmen beyond the sea. On behalf of my countrymen I thank you for your sympathetic message.

— William H. Taft"

Crossing the Bar

Sunset and evening star,
And one clear call for me.
And may there be no moaning of the bar,
When I put out to sea.

But such a tide as moving seems asleep,
Too full for sound and foam,
When that which drew from out the boundless deep
Turns again home.

Twilight and evening bell,
And after that the dark!
And may there be no sadness of farewell,
When I embark.

For tho' from out our Bourne of Time and Place
The flood may bear me far,
I hope to see my Pilot face to face
When I have crost the bar.

— Alfred Lord Tennyson

LIST OF THE DEAD

The following list of passengers missing from the *Titanic*, revised from last reports from the *Carpathia*, contains only 914 actual names out of the total of 1,635 lost, but many more are accounted for in the steerage reports under the word "family." Still more of the victims in the steerage have not yet been named, and few, if any, of the names of the missing among the crew have been reported.

FIRST CABIN

Anderson, Harry
Allison, H. J.
Allison, Mrs., and maid
Allison, Miss
Andrews, Thomas
Artagavoytia, Ramon
Astor. Col. J. J., and servant
Anderson, Walker
Beattie, T.
Brandies, E.
Mrs. Wm. Bucknell's maid
Baumann, J.
Baxter, Mr. and Mrs. Quigg
Bjornstorm, H.
Birnbaum, Jacob
Blackwell, S. W.
Borebank, J. J.
Bowden, Miss
Brady, John B.
Brewe, Arthur J.
Butt, Major A.
Clark, Walter M.
Clifford, George Q.
Colley, E. P.
Cardeza, T. D. M., servant of
Cardeza, Mrs. J. W., maid of
Carlson, Frank
Case, Howard B.
Cavendish, W. Tyrrell
Corran, F. M.
Corran, J. P.
Chafee, Mr. H. I.

Chisholm, Robert
Compton, A. T.
Crafton, John B.
Crosby, Edward G.
Cumings, J. Bradley
Davidson, Thornton
Dulles, William G.
Douglas, W. D.
Nurse of Douglas, Master R.
Eustis, Miss E. M. (may be reported saved as Miss Ellis)
Evans, Miss E.
Fortune, Mark.
Foreman, B. L.
Fortune, Charles
Franklin, T. P.
Futrelle, J.
Gee, Arthur
Goldenburg, E. L.
Goldschmidt, G. B.
Greenfield, G. B.
Giglio, Victor
Guggenheim, Benjamin Servant of Harper, Henry S.
Hays, Charles M.
Maid of Hays, Mrs. Charles M.
Head, Christopher
Hilliard, H. H.
Hopkins, W. F.
Hogenheim, Mrs. A.
Harris, Henry B.
Harp, Mr. and Mrs. Charles M.
Harp, Miss Margaret, and maid
Hoyt, W. F.

Holverson, A. M.
Isham, Miss A. E.
Servant of J. Bruce Ismay
Julian, H. F.
Jones, C. C.
Kent, Edward A.
Kenyon, Mr. and Mrs. F. R. (may be reported saved as Kenchen and Kennyman)
Kimball, Mr. and Mrs. E. N. (may be reported saved as Mr. and Mrs. E. Kimberley)
Klober, Herman
Lambert, Williams
Lawrence, Arthur
Long, Milton
Longley, Miss G. F.
Lewy, E. G.
Lindsholm, J. (may be reported saved as Mrs. Sigrid Lindstrom)
Loring, J. H.
Lingrey, Edward
Maguire, J. E.
McCaffrey, T.
McCaffrey, T., Jr.
McCarthy, T., Jr.
Marvin, D. W.
Middleton, J. C.
Millett, Frank D.
Minahan, Dr. and Mrs.
Marechal, Pierre
Meyer, Edgar J.
Molson, H. M.

Moore, C., servant
Natsch, Charles
Newell, Miss T.
Nicholson, A. S.
Ovies, S.
Ostby, E. C.
Ornout, Alfred T.
Parr, M. H. W.
Pears, Mr. and Mrs
  Thomas
Penasco, Mr. Victor
Partner, M. A.
Payne, V.
Pond, F., and maid
Porter, Walter
Reuchlin, J.
Maid of Robert, Mrs.
  E.
Roebling, W. A., 2d.
Rood, Hugh R.
Roes, J. Hugo
Maid of Countess
  Rothes
Rothschild, M.
Rowe, Arthur
Ryerson, A.
Shutes, Miss E. W.
  (probably reported
  saved as Miss
  Shutter)
Maid of Mrs. G. Stone
Straus, Mr. and Mrs.
  Isidor
Silvey, William B.
Maid of Mrs. D. C.
  Spedden
Speden, Master D., and
  nurse
Spencer, W. A.
Stead, W. T.
Stehli, Mr. and Mrs.
  Max Frolisher
Sutton, Frederick
Smart, John M.
Smith, Clinch.
Smith, R. W.
Stewart, A. A. (may
  be reported saved as
  Frederick Stewart)
Smith, L. P.
Taussig, Mrs. Emil
Maid of Mrs. Thayer
Thayer, John B.
Thorne, C.
Vanderhoof, Wyckoff
Walker, W. A.
Warren, F. M.
White, Percival A.
White, Richard F.

Widener, G. D., and
  servant
Widener, Harry
Wood, Mr. and Mrs.
  Frank P.
Weir, J.
Wick, George D.
Williams, Duane
Wright, George

SECOND CABIN

Abelson, Samson
Andrew, Frank
Ashby, John
Aldworth, C.
Andrew, Edgar
Beacken, James H.
Brown, Mrs.
Banfield, Fred
Beight, Nail
Braily, Bandsman
Breicoux, Bandsman
Bailey, Percy
Bainbridge, C. R.
Byles, the Rev. Thos.
Beauchamp, H. J.
Beesley, Lawrence
Berg, Miss E.
Benthan, I.
Bateman, Robert J.
Butler, Reginald
Botsford, Hull
Boweener, Solomon
Berriman, William
Clarke, Charles
Clark, Bandsman
Corey, Mrs.
Carter, Rev. Ernest
Carter, Mrs.
Coleridge, Reginald
Chapman, Charles
Cunningham, Alfred
Campbell, William
Collyer, Harvey
Corbett, Mrs. Irene
Chapman, John R.
Chapman, Mrs. E.
Colander, Erie
Cotterill, Harry
Charles, Wm. (probably
  reported saved as
  Wm. Charles)
Deacon, Percy
Davis, Charles (may be
  reported saved as

John Davies)
Debben, William
De Brits, Jose
Danborny, H.
Drew, James
Drew, Master M.
David, Master J. W.
Duran, Miss A.
Dounton, W. J.
Del Vario, S.
Del Vario, Mrs.
Enander, Ingvar
Eitmiller, G. F.
Frost, A.
Fynnery, Mr.
Faunthrope, H.
Fillbrook, C.
Funk, Annie
Fahlstrom, A.
Fox, Stanley, N.
Greenberg, S.
Giles, Ralph
Gaskell, Alfred
Gillespi, William
Gilbert, William
Gall, Harry
Gall, S.
Gill, John
Giles, Edgar
Giles, Fred
Gale, Harry
Gale, Phadruch
Garvey, Lawrence
Hickman, Leonard
Hickman, Lewis
Hume, bandsman
Hickman, Stanley
Hood, Ambrose
Hodges, Henry P.
Hart, Benjamin
Harris, Walter
Harper, John
Harper, Nina
Harbeck, W. H.
Hoffman, Mr.
Hoffman, child
Hoffman, child
Herman, Mrs. S.
Howard, B.
Howard, Mrs. E. T.
Hale, Reginald
Hamatainen, A., and
  infant son (probably
  reported saved as
  Anna Harnlin)
Hilunen, M.
Hunt, George
Jacobson, Mr.
Jacobson, Mrs.

Jacobson, Sydney
Jeffery, Clifford
Jeffery, Ernest.
Jenkin, Stephen
Jarvis, John D.
Keane, Daniel
Kirkland, Rev. C.
Karnes, Mrs. F. G.
Keynaldo, Miss
Krillner, J. H.
Krins, bandsman
Knight, R.
Karines, Mrs.
Kantar, Selna
Kantar, Mrs. (probably
  reported saved as
  Miriam Kanton)
Lengam, John
Levy, P. J.
Lahtigan, William
Lauch, Charles
Leyson, R. W. N.
Laroche, Joseph
Lamb, J. J.
McKane, Peter
Milling, Jacob
Mantville, Joseph
Malachard, Noll (may
  be reported saved as
  Mme. Melicard)
Moraweck, Dr.
Mangiovaccli, E.
McCrae, Arthur G.
McCrie, James M.
McKane, Peter D.
Mudd, Thomas
Mack, Mary
Marshall, Henry
Mayberg, Frank H.
Meyer, August
Myles, Thomas
Mitchell, Henry
Matthews, W. J.
Nessen, Israel
Nicholls, Joseph C.
Norman, Robert D.
Nasser, Nicholas (may
  be reported saved as
  Mrs. Nasser)
Otteo, Richard
Phillips, Robert
Ponesell, Martin (may
  be reported saved as
  M. F. Pososons)
Pain, Dr. Alfred
Parkes, Frank
Pengelly, F.
Pernot, Rene
Peruschitz, the Rev.

Parker, Clifford
Paulbaum, Frank
Rogers, Getina
  (probably reported
  saved as Miss E.
  Rogers)
Renouf, Peter E.
Rogers, Harry
Reeves, David
Slemen, R. J.
Sjoberg, Hayden
Slatter, Miss H. M.
Stanton, Ward
Sinkkonen, A.
  (probably reported
  saved as Anna
  Sinkkanea)
Sword, Hans K.
Stokes, Philip J.
Sharp, Percival
Sedgwick, Mr.
Smith, Augustus
Sweet, George
Sjostedt, Ernst
Toomey, Ellen (may be
  reported saved as
  Ellen Formery)
Taylor, bandsman
Turpin, William
Turpin, Mrs. Dorothy
Turner, John H.
Trouneansky, M.
Tervan, Mrs. A.
Trant, Mrs. Jessie
  (probably reported
  saved as Mrs. Jessie
  Traut)
Veale, James
Wilhelm, Chas
  (probably reported
  saved as Chas.
  Williams)
Watson, E.
Woodward, bandsman
Ware, William C.
Weiz, Leopold
Wheadon, Edward
Ware, John J.
Ware, Mrs. (may be
  reported saved as
  Miss F. Mare)
West, E. Arthur
Wheeler, Edwin
Wenman, Samuel

THIRD CLASS— STEERAGE

Allum, Owen
Alexander, William
Adams, J.
Alfred, Evan
Allen, William
Akar, Nourealain
Assad, Said
Alice, Agnes
Abbing, Anthony
Aks, Tilly
Attala, Malakka
Ayont, Bancura
Ahmed, Ali
Alhomaki, Ilmari
Ali, William
Anders, Gustafson
Assam, Ali
Asin, Adola
Anderson, Albert
Anderson, Ida
Anderson, Thor
Aronson, Ernest
Ahlin, Johanna
Anderson, Anders, and
  family
Anderson, Carl
Anderson, Samuel
Andressen, Paul
Augustan, Albert
Abelsett, Olai
Adelseth, Karen
Adolf, Humblin
Anderson, Erna
Angheloff, Minko
Arnold, Josef
Arnold, Josephine
Asplund, Johan
Braun, Lewis
Braun, Owen
Bowen, David
Beavan, W.
Bachini, Zabour
Belmentoy, Hassef
Badt, Mohamet
Beros, Yazbeck
Barry, —
Buckley, Katharine
Burke, Jeremiah
Barton, David
Brocklebank, William
Bostandyeff, Cuentche
Benson, John
Billiard, A., and two
  children

Bontos, Hanna
Baccos, Boulos
Bexrous, Tannous
Burke, John
Burke, Catharine
Burke, Mary
Burns, Mary
Berglind, Ivar
Balkie, Cerin
Brobek, Carl
Backstrom, Karl
Berglund, Hans
Bjorkland, Ernest
Can, Ernest
Crease, Ernest
Cohett, Gurshon
Coutts, Winnie, and
two children
Cribb, John
Cribb, Alice C.
Catavelas, Vassilios
Caram, Catharine
Cannavan, P.
Carr, Jenny
Chartens, David
Conline, Thomas
Celloti, Francesco
Christmann, Emil
Coxon, Daniel
Corn, Harry
Carver, A.
Cook, Jacob
Chip, Chang
Chanini, Georges
Chronopolous, D.
Connaghton, M.
Connors, P.
Carls, Anderson
Carlsson, August
Coelhe, Domingo
Carlson, Carl
Coleff, Sotie
Coleff, Peye
Cor, Ivan, and family
Calic, Manda
Calic, Peter
Cheskosic, Luka
Cacic, Gego
Cacic, Luka
Cacic, Taria
Carlson, Julius
Crescovic, Maria
Dugemin, Joseph
Dean, Bertram
Dorkings, Edward
Dennis, Samuel
Dennis, William
Drazenovic, Josef
Daher, Shedid

Daly, Eugene
Dwar, Frank
Davies, John
Dowdell, E.
Davison, Thomas
Davison, Mary
Dahl, Charles
Drapkin, Jennie
Donahue, Bert
Doyle, Ellen
Dwyer, Tillie
Dakic, Branko
Danoff, Yoto
Dantchoff, Christo
Denkoff, Mitto
Dintcheff, Valtcho
Dedalic, Regzo
Dahlberg, Gerda
Demossemacker, E.
Demossemacker, G.
Dimic, Jovan
Dahl, Mauritz
Dalbom, E., and family
Dyker, Adolph
Dyker, Elizabeth
Everett, Thomas
Empuel, Ethel
Elsbury, James
Elias, Joseph
Elias, Joseph
Elias, Hannah
Elias, Foofa
Emmet, Thomas
Ecimosic, Joso
Edwardson, Gustave
Eklund, Hans
Ekstrom, Johan
Ford, Arthur
Ford, M., and family
Franklin, Charles
Foo, Cheong
Farrell, James
Flynn, James
Flynn, John
Foley, Joseph
Foley, William
Finote, Lingi
Fischer, Eberhard
Goodwin, F., and
family
Goldsmith, F., and
family
Guest, Frank
Green, George
Garfirth, John
Gillinski, Leslie
Gheorgeff, Stano
Ghemat, Emar
Gerios, Youssef

Gerios, Assaf
Ghalil, Saal
Gallagher, Martin
Ganavan, Mary
Glinagh, Katie
Glynn, Mary
Gronnestad, Daniel
Gustafsch, Gideon
Goldsmith, Nathan
Goncalves, Mancel
Gustafson, Johan
Graf, Elin
Gustafson, Alfred
Hyman, Abraham
Harknett, Alice
Hane, Youssef, and two
children
Haggendon, Kate
Haggerty, Nora
Hart, Henry
Howard, May
Harmer, Abraham
Hachini, Najib
Helene, Eugene
Healy, Nora
Henery, Della
Hemming, Nora
Hansen, Claus
Hansen, Fanny
Heininan, Wendla
Hervonen, Helga, and
child
Haas, Alaisa
Hakkurainen, Elin
Hakkurainen, Pekka
Hankomen, Eluna
Hansen, Henry
Hendekovic, Ignaz
Hickkinen, Laina
Holm, John
Hadman, Oscar
Haglund, Conrad
Haglund, Ingvald
Henriksson, Jenny
Hillstrom, Hilda
Holten, Johan
Ing, Hen
Iemenen, Manta
Ilmakangas, Pista
Ilmakangas, Ida
Ilieff, Kriste
Ilieff, Ylio
Ivanhoff, Kanie
Johnson, A., and
family
Jamila, N., and child
Jenymin, Annie
Johnstone, W.
Joseph, Mary

Jeannasr, Hanna
Johannessen, Berdt
Johannessen, Elias
Johansen, Nils
Johanson, Oscar
Johansson, Gustav
Johkoff, Lazer
Johnson, E., and
family
Johnson, Jakob
Johnsson, Nils
Jansen, Carl
Jardin, Jose
Jensen, Hans
Johansson, Eric
Jussila, Eric
Jutel, Henry
Johnsson, Carl
Jusila, Katrina
Jusila, Maria
Keefe, Arthur
Kassen, Housseni
Karum, F., and child
Kelly, Anna
Kelly, James
Kennedy, John
Kerane, Andy
Kelley, James
Keeni, Fahim
Khalil, Lahia
Kiernan, Philip
Kiernan, John
Kilgannon, Theo
Kakic, Tido
Karajis, Milan
Karkson, Einar
Kalvig, Johannes
King, Vin., and family
Kallio, Nikolai
Karlson, Nils
Klasson, K., two
children
Lovell, John
Lob, William
Lobb, Cordelia
Lester, James
Lithman, Simon
Leonard, I.
Lemberopolous, P.
Lakarian, Orsen
Lane, Patrick
Lennon, Dennis
Lam, Ah
Lam, Len
Lang, Fang
Ling, Lee
Lockyer, Edward
Latife, Maria
Lennon, Mary

Linehan, Michael
Leinenen, Antti
Lindell, Edward
Lindell, Elin
Lindqvist, Vine
Larson, Viktor
Lefebre, F., and family
Lindblom, August
Lulic, Nicola
Lundal, Hans
Lundstrom, Jan
Lyntakoff, Stanke
Landegren, Aurora
Laitinen, Sotia
Larsson, Bengt
Lasson, Edward
Lindahl, Anna
Lundin, Olga
Moore, Leonard
Mackay, George
Meek, Annie
Mikalsen, Sander
Miles, Frank
Miles, Frederick
Morley, William
McNamee, Neal
McNamee, Ellen
Meanwell, Marian
Meo, Alfonso
Maisner, Simon
Murdlin, Joseph
Moore, Belle
Moor, Meier
Maria, Joseph
Mantour, Mousea
Moncarek, O., two
children
McElroy, Michael
McGowan, Katharine
McMahon, —
McMahon, Martin
Madigan, Maggie
Manion, Margaret
Mechan, John
Mocklare, Ellis
Moran, James
Mulvihill, Bertha
Murphy, Kate
Mikanen, John
Melkebuk, Philemon
Merms, Leon
Midtsjo, Carl
Myhrman, Oliver
Myster, Anna
Makinen, Kale
Mustafa, Nasr
Mike, Anna
Mustmans, Fatina
Martin, Johan

Malinoff, Nicola
McCoy, Bridget
Markoff, Martin
Marinko, Dimitri
Mineff, Ivan
Minkoff, Iazar
Mirko, Dika
Mitkoff, Nitto
Moen, Sigurd
Nancarror, William
Nomagh, Robert
Nakle, Trotik
Naked, Maria
Nosworthy, Richard
Naughton, Hannah
Norel, Manseur
Niels, —
Nillson, Herta
Nyoven, Johan.
Naidenoff, Penke
Nankoff, Minko
Nedelic, Petroff.
Nenkoff, Christie
Nilson, August
Nirva, Isak
Nandewalle, Nester
O'Brien, Dennis
O'Brien, Hanna
O'Brien, Thomas
O'Donnell, Patrick
Odele, Catherine
O'Connor, Patrick
O'Neill, Bridget
Olsen, Carl
Olsen, Ole
Olsen, Elin
Olsen, John
Ortin, Amin
Odahl, Martin
Olman, Velin
Olsen, Henry
Olman, Mara
Olsen, Elide
Orescovic, Teko.
Pedruzzi, Joseph
Perkin, John
Pearce, Ernest
Peacock, T., two
children
Potchett, George
Peterson, Marius
Peters, Katie
Paulsson, A., and
family
Panula, M., and family
Pekonami, E.
Peltomaki, Miheldi
Pacruic, Mate
Pacruic, Tamo

Pastche, Petroff
Pietcharsky, Vasil
Palovic, Vtefo
Petranec, Matilda
Person, Ernest
Pasic, Jacob
Planke, Jules
Peterson, Ellen
Peterson, Olaf
Peterson, Wohn
Rouse, Richard
Rush, Alfred
Rogers, William
Reynolds, Harold
Riordan, Hannah
Ryan, Edward
Rainch, Razi
Roufoul, Aposetun
Read, James
Robins, Alexander
Robins, Charity
Risian, Samuel
Risian, Emma
Runnestvet, Kristian
Randeff, Alexandre
Rintamaki, Matti
Rosblom, H., and
  family
Ridegain, Charles
Sadowitz, Harry
Saundercock, W.
Shellark, Frederick
Sage, Jno., and family
Sawyer, Frederick
Spinner, Henry
Shorney, Charles
Sarkis, Lahound
Sultani, Meme
Stankovic, Javan
Salini, Antoni
Seman, Betros
Sadlier, Matt
Scanlon, James
Shaughnessay, P.
Simmons, John
Serota, Maurice
Somerton, F.
Slocovski, Sleman
Sutchall, Henry
Sather, Simon
Storey, T.
Spector, Woolf
Sirayman, Peter
Samaan, Jouseef
Saiide, Barbara
Saad, Divo
Sarkis, Madiresian
Shine, Ellen
Sullivan, Bridget

Salander, Carl
Sepelelanaker, Alfons
Skog, Wm., and family
Solvang, Lena
Stranberg, Ida
Strilik, Ivan
Salonen, Ferner
Sivic, Husen
Svenson, Ola
Svedst, —
Sandman, Mohan
Saljilsvick, Anna
Schelp, Peter
Sihvola, Antti
Slabenoff, Peter
Staneff, Ivan
Stoytcho, Mikoff
Stoytenoff, Ilia
Sydcoff, Todor
Sandstrom, Agnes, and
  two children
Sheerlinch, Joan
Smiljanik, Mile
Strom, E., and child
Svensson, John
Swensson, Edwin
Tobin, Roger
Thomas, Alex
Theobald, Thomas
Tomlin, Ernest
Thorneycroft, P.
Thorneycroft, F.
Torber, Ernest
Trembisky, Berk
Tilley, Edward
Tamini, Hilion
Tannans, Daper
Thomas, John
Thomas, Charles
Thomas, Tannous
Tumin, T., and infant
Tikkanen, Juho
Tonglin, Gunner
Turoin, Stefan
Turgo, Anna
Tedoreff, Ialie
Usher, Haulmer
Nzelas, Jose
Vander and family
Vereruysse, Victor
Vjoblom, Anna
Vaciens, Adulle
Vandersteen, Leo
Vanimps, J., and
  family
Vatdevehde, Josep
Williams, Harry
Williams, Leslie
Ware, Frederick

Warren, Charles
Waika, Said
Wazli, Jousef
Wiseman, Philips
Werber, James
Windelor, Einar
Weller, Edward
Wennerstrom, August
Wendal, Olaf
Wistrom, Hans
Wiklund, Jacob
Wiklund, Carl
Wenzel, Zinhart
Wirz, Albert
Wittewrongel, Camille
Youssef, Brahim
Yalsevac, Ivan
Zakarian, Mapri
Zlevens, Rene
Zimmerman, Leo

All illustrations © Minalima. Pages 46; 49; 52; 60; 64; 89; 183; 201; 208; 219 © Corbis. Pages 118; 175; 195; 205; 271 © Getty Images.

Wreck and Sinking of the Titanic: The Ocean's Greatest Disaster
Copyright © 2012 by MinaLima

All rights reserved. No part of this book may be used or reproduced in any manner whatsoever without written permission except in the case of brief quotations embodied in critical articles and reviews. For information, address Harper Design, 10 East 53rd Street, New York, NY 10022.

HarperCollins books may be purchased for educational, business, or sales promotional use. For information, please write: Special Markets Department, HarperCollins*Publishers*, 10 East 53rd Street, New York, NY 10022.

First published in 2011 by:
Harper Design,
An Imprint of HarperCollins*Publishers*
10 East 53rd Street
New York, NY 10022
Tel: (212) 207–7000
Fax: (212) 207–7654
harperdesign@harpercollins.com
www.harpercollins.com

Distributed throughout the world by:
HarperCollins*Publishers*
10 East 53rd Street
New York, NY 10022
Fax: (212) 207–7654

ISBN: 978-0-06-206740-1
Library of Congress Control Number: 2011933894

Book design by Minalima

Printed in China
First Printing, 2012

Wreck and Sinking of the Titanic: The Ocean's Greatest Disaster was originally published in 1912. The text has been reproduced from the original book.

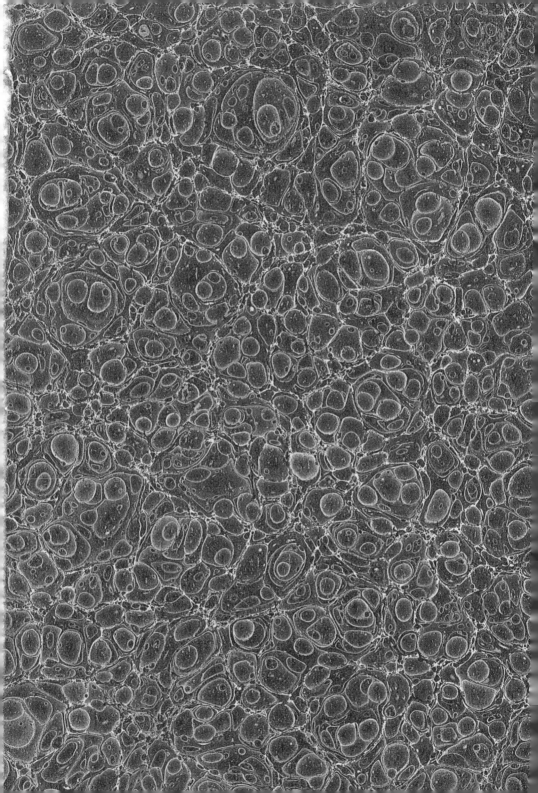

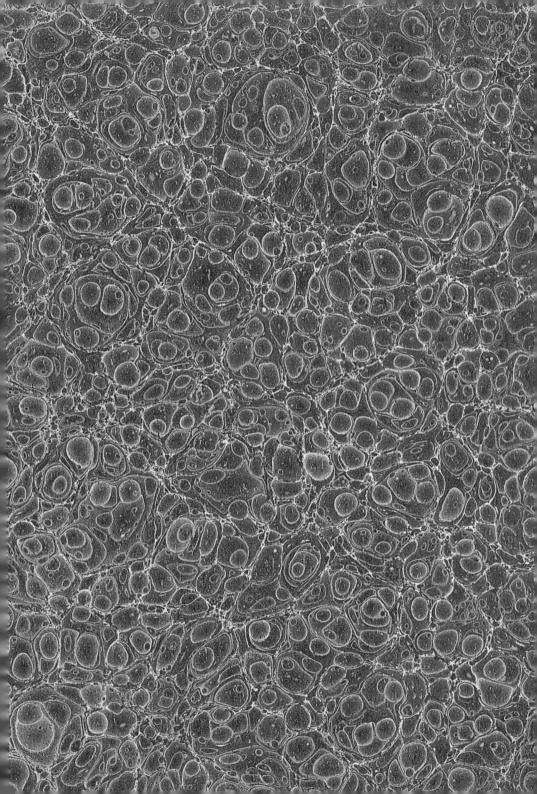